Replumbing the City

The publisher and the University of California Press Foundation gratefully acknowledge the generous support of the Ralph and Shirley Shapiro Endowment Fund in Environmental Studies.

Replumbing the City

WATER MANAGEMENT AS CLIMATE
ADAPTATION IN LOS ANGELES

Sayd Randle

UNIVERSITY OF CALIFORNIA PRESS

University of California Press
Oakland, California

Library of Congress Cataloging-in-Publication Data

Names: Randle, Sayd, author.
Title: Replumbing the city : water management as climate adaptation in Los
    Angeles / Sayd Randle.
Description: Oakland : University of California Press, [2025] | Includes
    bibliographical references and index.
Identifiers: LCCN 2024048575 (print) | LCCN 2024048576 (ebook) |
    ISBN 9780520394049 (cloth) | ISBN 9780520394056 (paperback) |
    ISBN 9780520394063 (epub)
Subjects: LCSH: Water-supply—California—Los Angeles—Management. |
    Climate change adaptation—California—Los Angeles. | Water conservation—
    California—Los Angeles. | Environmental policy—California—Los Angeles. |
    Urban ecology (Sociology)—California—Los Angeles.
Classification: LCC HD4464.L7 R37 2025 (print) | LCC HD4464.L7 (ebook) |
    DDC 333.91009794/93—dc23/eng/20250130
LC record available at https://lccn.loc.gov/2024048575
LC ebook record available at https://lccn.loc.gov/2024048576

GPSR Authorized Representative: Easy Access System Europe,
Mustamäe tee 50, 10621 Tallinn, Estonia, gpsr.requests@easproject.com

34 33 32 31 30 29 28 27 26 25
10 9 8 7 6 5 4 3 2 1

For my grandfather, Captain James P. Randle

Contents

Illustrations

Acknowledgments

This project has been shaped by many generous individuals who shared their time, space, and insights with me over the years, and to whom I am profoundly grateful. Born and raised on the East Coast of the United States, before beginning work on this project I'd only spent a handful of days in and around LA. I'm particularly indebted to the Angelenos I met during my time in the field: engineers, planners, NGO workers, environmental activists, RWAG members, greywater installers and adoptees, garden designers, researchers, and stormwater pilot project participants who welcomed me into their work and lives. Their kindness and openness made this book possible. Special thanks are also due to Tamar Walker and Wing and Wai Seto: to Wai and Wing for agreeing to rent a unit in their family's small apartment building to strangers and then doting on their new tenants with fresh-caught fish and long conversations throughout the years that followed, and to Tam for being the best possible roommate throughout my fieldwork. Wing passed away in late 2021 and is sorely missed by the people he cared for.

The long-term fieldwork that led to this book was financially supported by the Wenner Gren Foundation, the Yale Anthropology Department's Williams Fund, the Yale Institute for Biospheric Studies, and the

Yale School of the Environment's Hixon Center for Urban Sustainability (previously the Yale School of Forestry & Environmental Studies' Hixon Center for Urban Ecology). The Mellon Foundation and the American Council for Learned Societies provided time to complete the dissertation that preceded this manuscript and funding for follow-up fieldwork in 2018. Singapore Management University's College of Integrative Studies and Research Capacity Building Fund contributed financial support for developmental editing, mapmaking services, and indexing for the book.

Throughout graduate school I learned from many brilliant and caring individuals and communities. I owe particular gratitude to my dissertation committee. Shivi Sivaramakrishnan's rigor, clarity, and unwavering support shaped both this project and my understanding of how and why to pursue this kind of work. Michael Dove modeled extraordinary intellectual generosity, patience, and commitment to ethnographic thinking and practice. Karen Hébert's uncommon combination of critical acumen and personal warmth buoyed the work through dead ends and dark days. Laura Barraclough provided crucial guidance on thinking with the San Fernando Valley, and Erik Harms pushed me to continually question what, exactly, "the urban" entails. Over years of never-ending conversations, close readings, and potluck dinners (plus the occasional study beer), my cohort mates Samar Al-Bulushi, Nilay Erten, Sahana Ghosh, Ryan Jobson, Caroline Merrifield, and Gabriela Morales became key interlocutors and valued friends. Members of Yale's Dove Lab, Environmental Anthropology Collective, and Political Anthropology Working Group provided both fellowship and invaluable feedback. Many thanks to Aniket Aga, Lauren Baker, Samara Brock, Chandana Anusha, Matthew Archer, Jessica Barnes, Deepti Chatti, Annie Claus, Adrienne Cohen, Luisa Cortesi, Aysen Eren, Shaila Seshia Galvin, Radhika Govindrajan, Dana Graef, Amy Johnson, David Kneas, Myles Lennon, Atreyee Majumdar, Jessie Newman, Sarah Osterhoudt, Alyssa Paredes, Hosna Sheikholeslami, Jeff Stoike, Lily Zeng, and Amy Zhang for making those spaces so vibrant. Other essential New Haven comrades included George Bayuga, Heidi Lam, Max Lambert, Jessie Newman, Tri Phuong, Jacob Rinck, Chris Shughrue, and Katharine Walters. During those years I also came to know many wonderful scholars with resonant interests scattered across California, including Alyse Bertenthal, Emily Brooks, Jessica

Cattelino, Valerie Olson, Elizabeth Reddy, Caleb Scoville, and Julia Sizek, whose insights and generosity influenced this project's development both during and after my fieldwork in LA.

I had the immense good fortune to spend my immediate post-PhD years in California, thanks to a pair of postdoctoral fellowships. As a member of the Society of Fellows in the Humanities at the University of Southern California, I had the rare opportunity to start work on this manuscript while living in the same city where I undertook my fieldwork. Conversations with Mike Ananny, Jennifer Cool, Elizabeth Currid-Halkett, Janet Hoskins, Andy Lakoff, Nancy Lutkehaus, Jason Nguyen, Ashanti Shih, Sara Sligar, and Emily Zeamer were incredibly valuable as I began the work of transforming the dissertation into a book. Though deeply marked by the COVID pandemic, my time as a Ciriacy-Wantrup Postdoctoral Fellow at UC Berkeley was highly generative. Discussions with Nathan Sayre and members of Nancy Peluso's Land Lab helped many key ideas within the manuscript take shape. Special thanks to Ross Doll, Jared Finnegan, Hilary Faxon, Stephanie Postar, and Matthew Shutzer for making community over Zoom and on porches during those strange months. Leaving California for Texas at the end of 2021, I found myself in a remarkably welcoming environment at Texas A&M AgriLife's Urban Water Innovation and Sustainability Hub. I'm eternally grateful to Wendy Jepson for supporting my continued work on this project as I stepped into a role as a research scientist on her team.

To join Singapore Management University in early 2023, I moved half a world away from my previous life—the kind of distance that makes the excellent people I've found across Singapore even more of a gift. My colleagues at SMU have been incredibly supportive through the last months of writing and revision. Thanks to Jonathan Chase, Winston Chow, Maartje De Visser, Eric Fesselmeyer, Giovanni Ko, Andrew Koh, Elvin Lim, Darlene Machell, Teng Kuan Ng, Wen-Qing Ngoei, Yasmin Ortiga, Haesoo Park, Sovan Patra, Nona Pepito, Annika Rieger, Charlotte Setijadi, Ksenia Tatarchenko, Justin Tse, Terry Van Gevelt, Fiona Williamson, Aidan Wong, and Orlando Woods for kind words and smart advice as I wrestled the manuscript into its final form. Elsewhere on the island, Sahana Ghosh, Bhoomika Joshi, Canay Özden-Schilling, and Jacob Rinck have been wonderful writing partners, and the members of Singapore

Universities Network STS Works-in-Progress Group have been a superb source of fellowship and intellectual engagement.

I have shared sections of this manuscript in many forums over the years, and the thoughtful responses have shaped the form and argument of the book. I'm grateful for all those who engaged deeply with pieces of the work as copanelists, discussants, and audience members over the years. Special thanks are owed to those who offered feedback on draft chapters at various stages, including Samar Al-Bulushi, Alyse Bertenthal, Samara Brock, Emily Brooks, Deepti Chatti, Stephen Collier, Ross Doll, Hilary Faxon, Lyle Fearnley, Amelia Fiske, Maron Greenleaf, Lav Kanoi, Al Lim, Caroline Merrifield, Chandra Middleton, Canay Özden-Schilling, Tom Özden-Schilling, Haesoo Park, Jacob Rinck, Nathan Sayre, Shivi Sivaramakrishnan, and Shoko Yamada. Matthew Archer, Liviu Chelcea, Sahana Ghosh, and Gabriela Morales warrant special mention for generously and rigorously commenting on big chunks of the manuscript at critical junctures. Kathleen Kearns's developmental edit made the manuscript far clearer and more cohesive. Nick O'Gara was precise, rigorous, and unfailingly kind throughout his work creating the maps.

At UC Press, Naja Pulliam Collins has been an excellent editor and a pleasure to work with. Sylvie Bower provided great help in finalizing the manuscript. Thanks, too, to Stacy Eisenstark and Chloe Layman, who supported the project at earlier stages in the acquisitions and review process. I'm also extremely grateful to the anonymous peer reviewers for providing crucial insights and suggestions on the manuscript, feedback that markedly improved the work. Earlier versions of sections of the book were published in *American Anthropologist*, *City & Society*, and *Environment and Planning E: Nature and Space*. Thank you to the reviewers and editors of those publications for helping me develop the ideas presented throughout these pages.

To my family, thank you for encouraging my love of language, reading, and the environment from an early age, as well as for nurturing the sort of relentless curiosity that really can get tedious at times. To my parents, in particular: your unceasing support for my dreams has been foundational to this entire process, and I am indebted to you for such steadiness (and for so much else). My grandfather, Captain James P. Randle, has also always been a stalwart champion of my writing. He passed away at the age of ninety-six during my final weeks of work on the manuscript, and the book is dedicated to his memory. Finally, thank you to Ishan Basyal, my partner in all things, who has lived with this project since the day we met, talking me through ideas, reviewing drafts, and bolstering my spirits when finishing felt impossible. There's no one with whom I'd rather be building a life than you.

Introduction

Gesturing toward the enormous gravel pit that stretched before us, Ellen sighed appreciatively, brushing a mass of ash blonde hair from her eyes.[1] "I love thinking about how much water you could get into the aquifer here," she murmured to me and Nick. The three of us were crouched at the edge of the massive hole in the ground, a few hundred yards from the freeway in a sleepy section of the northeastern San Fernando Valley of Los Angeles (LA).[2] Beneath the rubble where we squatted sat the aquifer in question, a subterranean space formally known as the San Fernando Groundwater Basin.[3] Though the picturesque Verdugo Mountains loomed in the background, we gazed intently into the pit below us as Ellen described how a nearby flood control channel could be redirected to fill the depression with stormwater during downpours. From there, the water would be absorbed into the rocky, porous ground, eventually recharging the aquifer from which the city of LA draws a portion of its potable water supply.

Ellen, Nick, and I were spending a hazy August morning on a tour built around a vision for that groundwater basin's future. Ellen, the director of the environmental nongovernmental organization (NGO) where Nick worked and I conducted participant observation, acted as our guide, leading us through the northeastern Valley to help us appreciate the

1

landscape's unrealized ecological potential.[4] Her understanding of this section of the city was structured by the materiality of what sat beneath and flowed across its surface. Because of the area's rubbly soils and the groundwater basin lurking below them, Ellen saw the neighborhoods of the northeastern Valley as communities perched above LA's most egregiously underfilled water storage tank.[5] Well-schooled in local microclimates and hydrological patterns, she also understood this terrain as a sprawling floodplain for the huge volumes of runoff that crash down from the mountains lining its edges. Surveying the strip malls, auto wrecking lots, and subdivisions dominating this section of the city, she saw land that could be retrofitted to direct more of this stormwater into the aquifer. Put differently, to Ellen, the northeastern Valley was a space of urban ecological possibility, ground that—if properly managed—could capture and store new, vitally necessary water supplies for the city.

The idea for our excursion had been hatched weeks earlier when Ellen recounted a day spent leading an out-of-town fluvial geomorphologist and his graduate students around a cluster of current and potential stormwater recharge sites in the northeastern Valley. Cognizant of Ellen's deep knowledge of those neighborhoods, I asked if she might take me on a version of the outing. Though this landscape figured prominently in conversations around the office, as well as in our meetings with NGO and public agency collaborators, our time together in the northeastern Valley had almost all been spent at the organization's own stormwater capture pilot project. Nick admitted that visiting some of the other sites we spent so much time discussing sounded sort of fun and offered to chauffeur the group in his aging Prius. And so one not-yet-hot Sunday morning, we set off to see the pits, parks, flood control channels, and street medians that Ellen understood as key nodes (current and future) in LA's water provision system.

Ellen carried a printout of a LA Department of Water and Power (LADWP) map marked with major local stormwater infrastructures, but it quickly became apparent that the visual aid was more for our benefit than hers. While she had never lived in this part of the Valley, she operated from a detailed mental map of its established, under-construction, and prospective sites for water infiltration. When we departed the gravel pit, she directed us to a half-completed "green street" stormwater capture project, designed to transform the grassy strip between the road and the sidewalk into a series of basins to capture and infiltrate gutter water flows. A quarter of an hour later, we parked by a fence bearing the insignia of the LA County Department of Public Works and then passed through an unlocked gate to stare into a stormwater-spreading basin built in the middle of the last century.[6] Ellen's commentary on the sites seamlessly threaded explanations of the area's hydrology and landscape history with critiques of the development politics and hulking bureaucracies that she saw as impediments to realizing the northeastern Valley's stormwater capture potential.

Over tabbouleh and quesadillas at a counter service Lebanese-Mexican restaurant in a nearby strip mall, we speculated about all we'd seen. Ellen described the money (a lot) and agreements (many) it would probably take for the city to acquire and reengineer the gravel pit site, and Nick expressed mild surprise about how little of the green street project looked done. As it often did with us, the conversation eventually turned to Ellen's frustrated sense of urgency. California had been in a dry spell for years by that point. We drove under freeway signs reading "SERIOUS DROUGHT HELP SAVE WATER" on a daily basis. Storing more local stormwater within the city would help buffer LA from such recurrent bouts of aridity, and the floodplain of the northeastern Valley could infiltrate that reserve—but climate change was here, and so we needed to act *fast*, Ellen intoned. Chewing and nodding, I was struck, not for the first time, by the enormity of the work necessary to realize the potential Ellen saw in the urban landscape. The idea that the ground beneath our feet could provide such vital functions for the people of LA was enticing. But the degree of spatial transformation and sustained labor necessary for making that happen were, without question, intimidating.

Stories about water in LA tend to begin with the tale of an aqueduct. Since its 233-mile-long namesake pipeline began water deliveries in 1913, this city has been notorious for relying on a steady stream of liquid sourced (or stolen, depending on who's narrating) from distant landscapes.[7] Opening this book with scenes set at decidedly less prominent infrastructural sites—including some not yet built—within LA's borders, I am intentionally mirroring the transition that serves as my central object of analysis. This is an ethnography of efforts to transform metropolitan stormwater and wastewater flows into municipal water resources, centering the people and places entangled in these processes of urban retrofitting.

I approach these endeavors as a project of urban climate adaptation, interventions intended to sustain a functioning, populated city in the face of anticipated impacts of climate change. During my ethnographic field research, conducted between 2012 and 2018, I followed the work of local water agency employees, NGO staffers, environmental contractors, and residents striving to augment LA's municipal water supply with new, in-city water sources. Through an analysis of their ongoing efforts to redraw the boundaries of the city's sprawling system of water provision, an assemblage that stretches from the LADWP's downtown headquarters to the peaks of the Sierra Nevada and beyond, I show how a desire to maintain the condition of unlimited, uninterrupted water provision (to paying public utility customers) is haltingly reshaping this urban environment and its management.[8]

Producing new water supplies within the urban fabric entails a dramatic shift in the spatial logic of LA's approach to procuring this essential resource. The future availability of supplies from the lengthy pipelines that provide the vast majority of the city's water—around 90 percent at the time of this writing—is now widely understood as precarious.[9] The bulk of my research period coincided with California's 2012–2017 drought, a dry spell that everyone from environmental activists to city-employed water engineers to the aging techno star Moby (an invited speaker at a water conservation event hosted by LA's mayor in spring 2015) linked to a rapidly changing climate. During that period, politicians and climate scientists alike used images of a shrunken Sierra Nevada snowpack, the source of much of LA's piped-in water, to narrate the drought as a troubling preview of the weather patterns that will stress the city's water grid throughout the coming decades.[10] In casual conversations, mayoral directives, and water agency planning documents alike, the assumption that the distant snowbanks that had supplied the metropolis with water for a century were becoming less dependable was frequently invoked as a justification for initiatives intended to increase local water supply production, framed as a key step in enabling LA to withstand the altered climate.

These efforts to develop the city's diffuse urban flows as municipal water resources can be grouped into two categories: reuse and recharge.

Reuse projects seek to recycle wastewater currently directed to the Pacific Ocean via the sanitary sewer network. Recharge interventions—like those narrated by Ellen during our tour of the northeastern San Fernando Valley—aim to draw stormwater into aquifers sitting beneath the city, rather than allowing it to slide out to sea via the city's storm drains. All these initiatives seek to reconfigure the established routes by which water has long moved within the metropolis, a process I call *replumbing the city*. Corralling and redirecting such flows requires the development of new infrastructures and the retrofitting of old ones, expanding the footprint of LA's water provision system. But unlike hinterland reservoirs and pipelines that extend a city's imprint onto distant terrain, these are initiatives that refigure the water network's form and reach within the urban fabric itself.

Tracing such projects requires foregrounding landscapes and infrastructures that have not figured prominently in most accounts of LA and its water. While acknowledging the importance of the city's aqueducts and concrete-lined river, I focus primarily on the stormwater spreading grounds, sewage treatment plants, grassy power line easements, and domestic gardens and water fixtures that my interlocutors seek to embed in the municipal water grid in new ways.[11] As that list attests, developing the capacity to reuse and recharge water within the city entails interventions at many scales, ranging from intimate (like a bathroom sink) to massive (such as the city's 507 acres of stormwater percolation basins). While varied in terms of size and location, these projects are united by their connection to urban waters that since the city's early twentieth-century influx of imported supply have primarily been treated as wastes and hazards rather than as resources.

In addition to rerouting water, these emergent infrastructural arrangements also serve as conduits of urban spatial politics, contested sites where notions of expertise, duty, and state control are debated and refigured. Approaching these nodes as what anthropologist Antina Von Schnitzler terms *techno-political terrain*—that is, infrastructural sites where "central political questions of civic virtue, basic needs, and the rights and obligations of citizenship" are actively negotiated—allows me to track the emergent social, political, and material relations mediated by projects of urban reuse and recharge.[12] This understanding of infrastructure is aligned with

Map 1. City of LA's imported water sources. In 2023, LA relied on water sourced through the LA, California, and Colorado Aqueduct systems for roughly 90 percent of its municipal supply. Map by Nick O'Gara.

a growing body of anthropological scholarship that foregrounds infra-structure's relational nature and capacity to "assemble a range of spatial-ized relationships between political economic imperatives, technologies, natural processes, forms of sociality, social meanings, and modes of ritual action."[13] Building on such work, I start by recognizing infrastructures as sites where relations and flows can be arranged but rarely get permanently solidified—due in no small part to the unruly vitality of the people and materials involved in their ordering.[14]

The LA story diverges from most ethnographic accounts of urban infra-structure on a key register: many of the replumbing efforts—particularly those oriented toward recharge—seek to mobilize the capacity of other-than-human nature within the city for the purpose of redirecting flows of water. While new pipes, pumps, and filters are certainly involved in the project of drawing additional local water into LA's grid, so too are care-fully arranged soils, mulches, gravels, and plants in sections of the city overlying the San Fernando Basin. Proceeding from the notion that at-tentively managed slices of urban terrain can absorb and store new water supplies for LA, my interlocutors' efforts to produce such arrangements fit comfortably within the category that geographers Sara Nelson and Patrick Bigger call *infrastructural nature*, "policy approaches, scientific practices, discourses, and investment strategies that make ecosystems legible, governable, and investable as *systems of critical functions that sustain and secure (certain forms of) human life*."[15] Articulated in con-versation with ethnographic analyses of state- and NGO-led attempts to corral the capacities of plants, animals, and earth to modulate other ma-terial flows and circulations, this category captures the rapidly expanding array of landscapes across the globe now being interpreted and man-aged as service-providing natural infrastructures.[16] Melding landscape and techno-politics, many of these projects entail novel configurations of technology, capital, and labor within the targeted terrain. Reliant on diffuse inputs from natural systems themselves, these tend to be arrange-ments premised on the rationalization and partial commodification of the ecological functions in question.[17]

Examining this process of replumbing, I develop an analytic of *ad-aptation work* to parse how the labor of urban water management is being scripted, distributed, incited, discouraged, and valued within such infrastructural arrangements. Replumbing entails adaptation work to both reuse and recharge water supplies. Attending carefully to the diverse and differentiated forms of human exertion involved in the LA context, I elaborate the labor that my interlocutors put into producing and maintaining a novel suite of water-directing infrastructures to buffer the city from the impacts of climate change.[18] My approach builds on ethnographic scholarship grappling with the intersection of the lively materiality of infrastructural networks themselves and the sustained forms of material, political, and discursive work that makes resources move within those networks.[19] Some of this work involves the pipes and concrete and shovels commonly associated with projects of urban retrofitting. But some engages other tools and takes place in settings removed from the water-routing infrastructures themselves. Finicky computer models, contentious stakeholder meetings, tedious grant applications, promotional Instagram posts, and stubborn weeds are also important elements of these efforts. Considering such a range of materials and exertions reveals the extent of human time and energy this approach to urban climate adaptation demands.

Replumbing thus emerges as a project contingent on the efforts of actors not previously understood as involved in urban water supply production, a notable shift in resource and spatial governance. During the twentieth century, workers employed by public agencies carried out almost all the labor of LA's water provision, building and managing networks of massive aqueducts, reservoirs, pipes, pumps, and treatment plants. While some of the interventions I trace could be described in similar terms, others rely heavily on the efforts of NGO advocates and ultimately seek to enroll urban residents into long-term collaborations with soil and plants and gutters and sometimes clothes-washing machines to reroute water.[20] As I show, this expansion can offer openings for a range of actors to contest the roles of urban land, the state, and residents within arrangements of water provision, challenging established logics of control and forms of expertise within the water grid.

Such a shift in the water provision system's boundaries also draws the inputs of other-than-human nature into the water network—and likewise, into my analytical frame. To account for this expansion, the book approaches the work of water management as an ongoing collaboration between people, the built environment, and other-than-human nature, a form of what political theorist Alyssa Battistoni terms *hybrid labor*: "a collective, distributed undertaking of humans and nonhumans acting to reproduce, regenerate, and renew a common world."[21] Elaborating this concept, Battistoni argues that such a framing provides a way to grapple with the forms of other-than-human agency and relationality that mark such collaborations, as well as with unsettled questions of how value should be assigned within them—an orientation I aim to develop ethnographically through the chapters that follow.[22]

In interventions that mobilize the landscape's capacities in this manner, human imagination and effort are necessarily entangled with other-than-human nature. While attentive to the potential pitfalls of expanding the categories of work and labor beyond human exertions—particularly of what anthropologists Sarah Besky and Alex Blanchette have termed the risk of projecting "capital's fixation on the value of human labor onto all of the planet's energies"—I contend that approaching LA's replumbing through this lens clarifies a critical aspect of such projects.[23] Glossing such hybrid arrangements as a form of work enables an inclusive accounting of their constitutive inputs and energies without unduly flattening the forms of agency involved. Though resonant with recent works that center the liveliness, recalcitrance, and generativity of plants and animals in the production of urban space and politics, my emphasis here diverges subtly, drawing questions of value and remuneration squarely into considerations of such formations through the adaptation work analytic.[24] Conceptualizing the (strategically retrofitted) metropolitan landscape's capacity to reroute flows in this manner allows me to situate such contributions within the broader field of labor involved in replumbing the city.

Although it is grounded in the particularities of the LA case, my analysis is also relevant to understanding the profusion of so-called nature-based solutions featured in recent plans for urban resilience and sustainability across the globe. It highlights a concerning dynamic that often marks these well-meaning efforts: they can redistribute the work of urban resource management in a highly uneven manner, producing new forms of metropolitan spatial and social differentiation in the process. While these interventions seek to secure water for an entire public municipal provision network, they will not affect all areas of a city's landscape equally. As my tour of the northeastern San Fernando Valley with Ellen and Nick suggests, in LA water agency and NGO plans disproportionately target specific aquifer-connected terrain within the city for these projects. When managed for this function, that ground can provide far more water-capture benefits than can other areas within LA and thus draws more state attention, capital investment, and infrastructural installations to realize these ecological services than do other sections of the metropolis. Highlighting such a lopsided distribution within the urban fabric helps to reframe and rescale ascendant discourses of the capacity of nature-based infrastructures to bring resilience to "the city" more generally, drawing attention to the forms of intra-urban spatial difference that can emerge through such projects.

The stakes of such unevenness become clearer when one considers the histories of state neglect and toxic burdens that mark the neighborhoods targeted for these interventions. As I discuss at length in chapter 3, the rubbly floodplain of the northeastern Valley has long functioned as a racialized, pollutable periphery within the city's borders. Prohibited from purchasing land in other sections of LA by racist zoning laws, Black, Chicano, and Japanese American Angelenos flocked to the land during the first half of the twentieth century. Mirroring the racial capitalist dynamics recounted in other US cities, the second half of the century saw the proliferation of noxious industries within and the sluggish expansion of basic public infrastructure to the area's majority-nonwhite neighborhoods.[25] Today, this is a Latinx-majority area that remains economically marginalized and continues to grapple with both occasional floodwaters and the toxic traces of now-shuttered factories in its air and soils.[26]

Prioritizing the northeastern Valley for infrastructural nature interventions thus represents a notable increase of state and environmentalist attention to and investment in this landscape—a reversal that, some have suggested, is long overdue. Examining these projects as adaptation work, however, highlights a more ambivalent aspect of this approach. Here, both the work of nature and much of the human labor involved are scripted as inputs that can be had cheaply. While public agencies and NGOs (with some struggle) will fund the construction of distributed, nature-based infrastructures that absorb and percolate stormwater flows, the work of maintaining them is rarely accounted for in long-term plans and budgets.

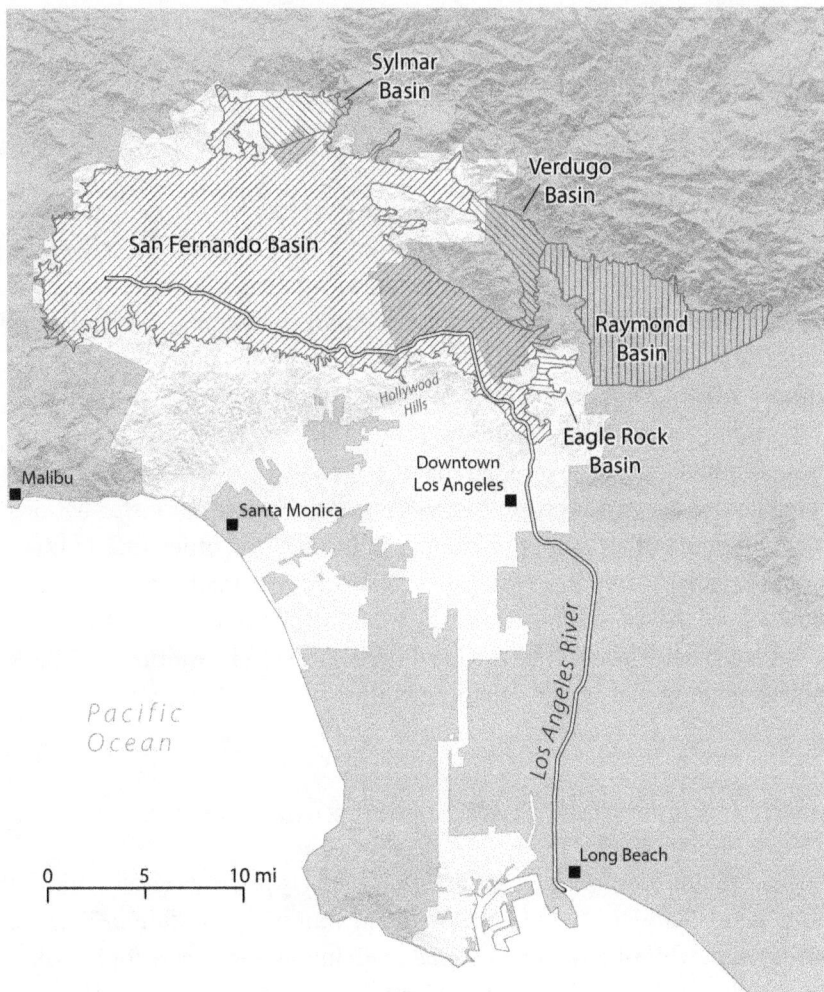

Map 2. San Fernando Valley Basin's four sub-basins, San Fernando, Sylmar, Verdugo, and Eagle Rock (as well as the adjacent Raymond Basin), superimposed onto the city of LA (depicted in white). The opportunities for stormwater recharge discussed throughout the book are concentrated near the eastern edge of the San Fernando Basin. Map by Nick O'Gara.

Instead, residents and community groups based in these neighborhoods—seldom the instigators of the projects—are assumed to be the appropriate, uncompensated stewards of the new array of stormwater infrastructures, tasked with the long-term labor of trash picking, degunking, weeding, and pruning to keep the installations functional. Premised on mobilizing the landscapes and communities of the northeastern Valley to collaborate in the production of new water resources for all of LA, these interventions thus emerge as both localized opportunities and localized obligations.

Through a sympathetic critique of such arrangements, the book uses Angelenos' efforts to replumb the city to highlight the unsettled, spatially uneven relationship between work and value at the core of these and so many other climate adaptation projects.[27] Like climate mitigation, adaptation entails a capacious remaking of established infrastructures, cultivated landscapes, and the built environment writ large—transformations that will demand an immense amount of human and other-than-human labor to realize. Attending closely to such work, I contend, reveals the new benefits, relations, and burdens that enacting adaptation entails. It also offers a crucial window into the lived dynamics and inequities that mark efforts to adapt to a climate-changed world.

MAKING A CITY OF DRAINS AND PIPELINES

About five minutes into the movie *Chinatown* (1974), the viewer joins the film's main character at a public hearing in a municipal building. There, a worked-up, white-haired speaker makes an impassioned case for building an aqueduct to bolster LA's water supply. "Los Angeles is a desert community," he tells the assembled onlookers. "Beneath these buildings, beneath the streets, is a desert. And without water, the dust will rise up and cover us as if we never existed."

Over the years, several of my key interlocutors have told me that they hate this line and blame it for planting an inaccurate idea about coastal Southern California's climate deep in people's minds, due to its prominent placement within a film that so many treat as the "true" story of water development in the region.[28] Setting aside the question of *Chinatown*'s cultural influence, they are unquestionably correct on the point about the local climate: LA is not located within a desert. Part of the unceded home-lands of the Gabrielino-Tongva Indian Tribe and the Fernandeño Tataviam Band of Mission Indians, the area's meandering river and lush riparian bottomlands drew Spanish colonizers in the late eighteenth century.[29] Averaging just under fifteen inches of rain per year, the city's climate is classified as Mediterranean. As is typical of areas bearing this label, LA has highly seasonal and extremely inconsistent precipitation. Rain rarely falls outside of the October to April window, and the total volume of precipitation shifts wildly from year to year—an inconstant rhythm common across much of California. In a memorable riff, the late journalist Marc Reisner once suggested that LA's characteristic swings between drought and deluge meant that "evolution, left to its own devices, might have created in a million more years the ideal creature for the habitat: a camel with gills."[30]

Given the variability of the local hydrology, the history of LA's water system over the past century features water managers and political leaders seeking to solve not just a volumetric problem but also a temporal one. Since the early twentieth century, enormous publicly funded programs of water development have smoothed the experience of this hydrological inconsistency across Southern California. That leveling out began with the construction of LA's eponymous aqueducts, which the LADWP owns and manages.[31] The first pipeline, completed in 1913, taps flows of Sierra Nevada snowmelt within the rural Owens Valley (known as Payahuunadü to its Native Paiute residents), located in northeastern California.[32] Unlimited, universal water provision for paying customers quickly became the norm within LA's city limits, an arrangement sustained with that aqueduct's imported water. As the century unfolded and the city continued to expand, LA lengthened its original pipeline to tap watersheds feeding Mono Lake in 1941 and completed a second, parallel aqueduct to transport more water from the same rural landscapes in 1970. To secure access to another set of distant sources, in 1928 the city joined neighboring jurisdictions to establish the Metropolitan Water District (MWD), a regional water wholesaler, which constructed the infrastructure necessary to pipe water from the Colorado River to Southern California in the late 1930s. Metropolitan later contracted with the newly built State Water Project to provide the Southland with still further supply from additional Northern California watersheds from the early 1970s onward. As a result, while there may be no local rainfall for half the year, potable water has flowed steadily through LA's water network to the city's four million residents.

During these decades of intensive hinterland water resource development, public agencies also constructed an extensive array of in-city infrastructures to corral local storm runoff. But the primary purpose of this network of concrete storm drains was flood prevention rather than water supply augmentation. Approaching stormwater as a threat to rapid residential and industrial growth within the floodplain of the LA River and its tributaries, twentieth-century engineers built out a skein of conduits to direct those flows away from valuable real estate and out to the Pacific Ocean as efficiently as possible. Likewise, over the course of the 1900s, the city constructed a quartet of wastewater treatment plants to cleanse the flows from LA's sanitary sewer network before sending the effluent on toward the sea. Though certainly not flawless—for instance, chronic underinvestment in the oceanside Hyperion Treatment Plant during the 1970s and 1980s led to persistent water quality issues off the city's coast—LA's infrastructures of waste and stormwater disposal have largely functioned as designed, serving to contain and direct flows understood as dangerous out of the city.

A reappraisal of the waters within these conduits is central to the replumbing efforts chronicled here. While these flows are still understood as hazardous and polluting in many circumstances, water managers now approach them as potential water resources, available for use if the city can find a way to effectively reroute and (in some cases) cleanse them. The growing appreciation of these substances' dual character is based in part on the range of regulatory, political, and environmental constraints that have circumscribed the city's access to its distant resources in recent decades.[33] Such shifts have been especially noticeable within the LA Aqueduct system. After a 1977 study revealed the ecosystem of Mono Lake to be in a state of near-collapse, environmental groups sued the LADWP for damage the aqueducts' water diversions had caused. A landmark 1983 ruling by the California Supreme Court found that the state has an obligation to protect beautiful and ecologically sensitive places like Mono Lake "as far as feasible." Following further litigation, in 1994 the State Water Resources Control Board substantially reduced the amount of water LA could draw from Mono Lake tributaries. Not long after this decision, the

Figure 1. Channelized LA River near downtown LA, transporting a stream of urban runoff mixed with wastewater treatment plant effluent. Photo by author.

Great Basin Unified Air Pollution Control District began a round of wrangling with the LADWP over the hazardous particulate matter emanating from the sucked-dry bed of Owens Lake. The long-awaited 2014 settlement of these disputes substantially increased the volume of water that the city was required to leave in the Owens Valley for the purposes of dust mitigation, further increasing LA's reliance on water purchased from Metropolitan during the 2012–2017 drought.

Notably, while my interlocutors often discussed that protracted dry spell as a worrying preview of hydrological conditions to come under climate change, they described the historically wet winter of 2017 in similar terms. That drought-ending season was marked by so much precipitation that evacuation orders were issued for over one hundred thousand residents downstream from one of California's largest dams. Regional climate models predict a near future in which the state's characteristic swings between parched and soaked become even more dramatic. Droughts will be longer and more intense.[34] But when precipitation finally arrives, a larger proportion will fall as rain, and the high-mountain snowbanks will melt faster than in the past, compromising lower-elevation water storage arrangements.[35] This accumulation of threats to LA's ability to draw water from its distant sources has bolstered the growing managerial consensus that sustaining a reliable water system requires the development of previously untapped urban flows. In light of this view, the city seeks to protect the dependability of its consumer-facing delivery network from moments when the older sources run dry or become otherwise inaccessible.

As noted, these interventions target two distinct sets of flows: wastewater and stormwater. Those focused on wastewater seek to recycle sanitary sewer effluent, either via centralized advanced treatment plants that would reincorporate the cleansed flows into the water grid or by simpler, building-scale systems that would enable reuse on-site (as detailed in chapter 2, these scales of recycling are associated with distinct communities of practice). Stormwater recharge projects, concentrated in the northeastern corner of the city, aim to corral diffuse flows of runoff currently destined for the city's storm drain system into land-based infrastructures that facilitate its percolation into the San Fernando Groundwater Basin. The LADWP's most recent Urban Water Management Plan (UWMP) projects that the annual volume of water sourced from the combination of stormwater, groundwater, and recycled water will roughly triple from its 2016–2020 average of 54,712 acre-feet by fiscal year 2044–2045, climbing to 163,415 acre-feet and representing 23 percent of the city's supply mix.[36] While certainly not severing the city's ties to its hinterland sources, if realized this program of replumbing would substantially expand LA's dependence on waters supplied within city borders.

CLIMATE ADAPTATION AND THE PRODUCTION OF URBAN SPACE

Seated across a small conference table from Evan, I waited while he considered my question about what, at worst, LA's water system circa 2050 could look like. A white man and an early-career engineer employed by a local public agency, Evan had agreed to sit for an interview with me after a year of crossing paths at outreach meetings related to a stormwater planning study that he was helping manage. He had spent a considerable chunk of our conversation explaining how his team had incorporated climate data into its hydrological model of the LA geological basin to help ensure that their planning framework wasn't based on outdated assumptions about local precipitation patterns. While acknowledging the limitations of the available information, Evan had told me he believed that, thanks to the help of climate scientists, his team had been able to build a model that reflected the longer droughts and extraordinarily wet years those experts anticipated for the region. Based on this experience, he predicted that local water agencies would be shaping their infrastructure and policy around climate change in the years to come, adjusting operations to capture as much water as possible during local storms.

Given his optimism about this trajectory, I wasn't shocked that Evan's version of a pessimistic arc for the city was light on the disaster imagery, a sharp contrast to well-known accounts of future LA as a desiccated semi-hellscape.[37] In fact, his response quickly took a hopeful turn. "Worst case, what comes to mind: no imported water supply," he told me. "Though even that worst case could become a best case as well." Continuing, he explained that if distant water sources shrank due to climate change, it would put a greater burden on public agencies to innovate further and capture more water locally. With an all-of-the-above approach to new infrastructure development, Evan concluded, we could keep the system reliable, preventing floods and providing drinking water for everyone.

Such a response, while acknowledging climate change as real and powerful, presents it as a reasonably knowable and ultimately manageable phenomenon. It's an environmental future that assumes plenty of stasis: the LA of 2050 might not look exactly like the city we knew that day in late 2015, but it could be sustained as a recognizable version of that metropolis. In such framings, common among my public-agency-based interlocutors, urban climate adaptation emerges as a techno-managerial project of maintaining established socio-ecological relations by retrofitting the metropolitan fabric. As I explore elsewhere, more fundamentally disrupted visions of LA's future waterscape were also circulating widely during my fieldwork period.[38] My interlocutors whose professional work

did not directly address local water politics tended to sketch more apoc-
alyptic landscapes than Evan did when asked about the city's worst-case
water outcomes, echoing concurrent cultural production focused on systems of water provision in the US West.[39] Considering more technocratic
perspectives, largely drawn from my time among water managers, I elaborate the spatial relations produced by projects grounded in their com-
paratively mild assumptions about the degree of systemic transformation
necessary to sustain the city's water network.

In contrast to climate change mitigation, widely understood to demand
a dramatic transformation of hydrocarbon-dependent economies and
infrastructural arrangements to prevent the worst impacts of an altered
climate, climate adaptation is about making the unavoidable effects of climate change more survivable. As anthropologist Gökçe Günel succinctly
puts it, adaptation efforts concentrate "on managing the already observed
impacts of climate change as well as those predicted to happen with a high
degree of certainty."[40] While ethnographers have highlighted the contingent, often political effort that producing and interpreting climate models
entails, technocratic actors in many contexts now approach such predictions for localized climate change effects as actionable knowledge.[41] However, questions of what adaptive actions, exactly, should be taken in a given
setting tend to remain opaque and contested, matters of ongoing techno-
political wrangling.[42] Though diverse and uncertain, most climate adaptation efforts pursued by public agencies and NGO actors fit comfortably
within Günel's category of *technical adjustments*—that is, infrastructural
interventions that seek to enable the extension of established lifestyles and
power relations.[43] Unlike approaches that cast contemporary landscapes
as already ruined and in need of capacious, diverse projects of living differently within them, technocratic adaptation tends to assume that the status
quo is generally worth preserving—and largely preservable.[44]

Water management is inherently anticipatory work, necessarily
grounded in assumptions about environmental futures. As such, folding
expectations of a changing climate into city water planning represents a
shift in, rather than a wholesale break from, established managerial practices. In the LA context, adapting to an altered climate is a project of securing the water provision system from a familiar set of threats—only
magnified. As noted earlier, drought and deluge are both recurrent events

in the region. The storied floods of 1862, 1889, 1914, 1934, and 1938 all took lives and destroyed property across the LA basin, and the droughts of 1928–1935, 1976–1977, and 1987–1992 strained the city's water provision system. The goal of ensuring that such perturbations did not disturb the functioning of critical infrastructural systems has shaped the region's institutions, practices, and investments in water management for decades. Technocratic climate adaptation efforts, here and elsewhere, entail plenty of continuity with past projects—a dynamic I explore at some length in chapters 1, 2, and 3.

Yet while attending carefully to the arrangements and relations preserved by climate adaptation initiatives, I foreground the novel landscapes and forms of space, politics, and governance that they are producing within and beyond LA's borders. Evan's bullish account of the city's worst-case 2050 waterscape hints at these shifts. The imported water sources that he feared would dissipate—a heightened version of climate modelers' current predictions for their fate—are drawn from hinterland landscapes marked by the city's managerial presence since the early twentieth century. Building the LA Aqueduct was enabled by city agents buying up land and water rights throughout remote areas in northeastern California. LA retains these ownership claims, enabling the city to circumscribe development and resource access in local Indigenous and settler communities.[45] While decidedly rural, these are best understood as landscapes and communities shaped by the material and political imprints of the city's ongoing extractions."[46] A shrinking snowpack would reduce the flows within these rural areas, a decline that would affect both local environments and LA's ability to siphon water to the city.

As Evan suggested, such a trajectory could spur agencies within LA to intensify their efforts to capture local stormwater and wastewater flows. This transition would obviously represent a spatial adjustment to LA's urban metabolism, reducing the draw from one landscape and incorporating new flows from another.[47] Highlighting LA's protracted role in shaping hinterland terrain via water production helps to clarify the reality that managing land for this purpose can bring a wide range of impacts—social, political, and material—to the new areas conceived as potential infrastructural nature. Such effects are well documented in scholarship on extended urban water and electricity provision networks across the globe,

Figure 2. Fenced-in, open-channel stretch of the LA Aqueduct in the Owens Valley, roughly two hundred miles from the LADWP's downtown headquarters. Photo by author.

which have emphasized the uneven power dynamics and extractive political economic forms that accrue within rural, resource-providing communities.[48] These accounts also demonstrate the forms of human resistance and other-than-human vitality that frequently undermine such projects. Tracking an urban network's transformation driven by a contrasting spatial logic, I offer a portrait of the complex, sometimes contradictory relations that emerge when water provision projects "refigure the [space of the] city as its own solution."[49]

URBAN TERRAIN AS SERVICE-PROVIDING NATURE

The notion that land can be managed to provide a particular cluster of desired environmental functions is not a new one, particularly in the US West. Since the late nineteenth century, government-led projects have sought to maximize the hydrological potential of critical watersheds through targeted rural forest management efforts.[50] The more recent rise of the so-called ecosystem services paradigm, which seeks to measure and assign economic value to such contributions from the landscape, has played a central role in expanding and operationalizing this understanding of other-than-human nature.[51] In this context, the term *ecosystem services* is understood to denote the ecological benefits humans derive from an ecosystem. Clean water, produced through plants and soils filtering out pollutants, and breathable air, purified through sylvan respiration, are frequently cited as examples of such desirable functions and outputs. This framework has gained traction among economists, investors, and conservation practitioners since the 1970s, emerging in response to economists' dissatisfaction with the failure of neoclassical economics to assign market value to nature's contributions to the economy.[52] Approaching the non-human world as a collection of functions rather than as an inert stock of so-called natural capital, the ecosystem services approach emphasizes dynamism and complexity.[53]

Since the 1990s, when a landmark paper in the field of environmental economics estimated the cumulative value of the biosphere's ecosystem services at between $15 and $63 trillion, the growing desire to "protect" nonhuman nature by drawing its work into circuits of capital has spurred a range of programs of ecological measurement, pricing, and marketization.[54] As I detail in chapter 4, the work of making such services visible in processes of valuation remains fragmented and piecemeal in practice. But the notion underlying such projects—that slices of land can be managed to provide valuable ecological functions (rather than just commodifiable goods) for people—is increasingly mainstream, particularly in projects pursued in the name of climate resilience.[55]

Most people passing through LA's endless blocks of densely clustered shopping plazas and apartment complexes do not initially encounter the space as "nature," let alone land rife with potential ecosystem services. Across much of the city, the landscape is dominated by concrete and weathered storefronts, with spindly, non-native palm trees and the ubiquitous Starbucks mermaid logo providing much of the local greenery—a far cry from Western tropes of pristine wilderness or carefully tended countryside.[56] But while such popular understandings of urbanized space persist, a view of cities as urban ecosystems that might be managed to maximize human and ecological well-being has gained traction among policy actors and funding agencies across the globe over the past three decades.[57]

This is very much the case in contemporary LA, where rerouting water's movement through the city is often framed as constitutive of effective urban ecosystem management. For instance, in some environmentalist circles, domestic gardens are discussed not only as slices of land where resource (particularly water) consumption should be moderated, but as ground that should be cultivated to enhance local biodiversity, pollinator habitat, and food production—ends that, some contend, can be most effectively realized by redirecting a home's wastewater flows. And as my tour of potential stormwater infrastructure in the northeastern Valley suggests, environmentalists and engineers alike understand sections of the urban landscape as terrain that can and should be managed to maximize its ability to absorb rainwater and recharge local aquifers.[58] Such accounts of city land's potential functions are grounded in a critique of the twentieth-century approach to urban development that made LA into a place where both stormwater and wastewater are efficiently routed to the Pacific Ocean. Notably, most of these projects do not entail land changing hands but rely on retrofitting small, carefully selected sections of the urban environment on parcels the city already owns, like public parks, power line easements, street medians, and parkway strips. Private land, including residential properties, is also targeted for retrofitting, but in those cases the current property owners are generally assumed to be the appropriate developers of these new infrastructures.

Within and beyond LA, many of the retrofits I examine are known as *green infrastructure* or *nature-based solutions*. Though the scale and form of these installations can vary, they are all generally designed to collect and hold water and let it soak into the ground. Proponents celebrate these infrastructures for the wide range of functions that they provide the urban environment. Such nature-based solutions, I was frequently told during my fieldwork, reduce flood risk, improve local water quality, increase urban biodiversity, sequester carbon, and replenish subsurface groundwater basins. My NGO interlocutors were usually the loudest cheerleaders for this expansive understanding of these ecosystem services, not to mention a range of associated social and economic benefits. During presentations at community forums and working group meetings with city agency workers, I heard them chastise other attendees for framing the contributions of these installations too narrowly, underselling the value that they

Figure 3. Two volunteers planting a newly dug parkway basin in front of a residential property. Photo by Samantha Bode.

provide to the city. In contrast, city water managers tended to cite a relatively circumscribed range of environmental functions associated with the projects (biodiversity and carbon sequestration rarely surfaced). But their preferred terminology for the installations—"multi-benefit projects"—signals the range of services the engineers believe these infrastructures can bring to the urban landscape.

As such framings suggest, the replumbing efforts I trace (projects of recharge and reuse alike) were understood by their proponents as interventions that could improve local environments and quality of life, in addition to bolstering municipal water supply. Following Caterina Scaramelli, I approach these attempts to reshape the urban landscape as a moral ecological undertaking—that is, a form of "ecological practice and thought in which morality—the concern with what is of value in life—is at stake."[59] Many of my interlocutors saw their efforts to transform the metropolitan space as a decidedly more-than-technical undertaking, one with the potential to address a range of long-standing social and ecological harms (within and beyond LA's borders) and buffer the city from the effects of climate change. Yet as Scaramelli has shown, a grounding in care or ethical commitments to future generations does not preclude moral ecological projects from producing burdens or marginalization for certain residents—a key dynamic in this context.

In the pages to come, I detail the roadblocks on the path from recognizing to realizing the urban landscape's ecological potential—that is, the often-frustrating work of cultivating this form of nature's labor. As Ellen's impatience over our post-tour quesadillas suggests, green infrastructure proponents, including those situated within public agencies, often struggled to fund and develop these water-capture projects at a pace sufficient to substantively reroute the flow of runoff within LA. Though key decision-makers and funders increasingly acknowledge such projects as retrofits that could bolster the city's water supply, making the landscape's potential infusions of water legible to these actors was understood as a slippery, uncertain undertaking, which I explore in some detail in chapters 3 and 4. Likewise, this form of urban infrastructural nature retains a tenuous connection to processes of monetary valuation, an arrangement that complicates the politics of developing these projects within the space of the city, a core concern of chapters 4 and 5. Elaborating such stuttering progress toward making money flow to these projects points to contradictions inherent in techno-managerial efforts to realize a functional landscape of infrastructural nature within the city. It also clarifies the diverse forms of human exertion necessary to realizing the "services" that so many contemporary plans and projects seek to cultivate within the urban environment.

RETHINKING ENVIRONMENTAL JUSTICE THROUGH ECOSYSTEM DUTIES

Ariana welcomed me into her northeastern San Fernando Valley home one windy weekday evening in late autumn. A twenty-six-year-old Latina mother of two, she told me she needed to get her kids to bed soon and asked if we could keep our interview brief. There to learn about her experiences participating in an NGO-led pilot project that installed rainwater-absorbing green infrastructure on residential properties, I agreed, and we moved through my planned questions efficiently. Ariana quickly listed the changes to her home landscape, describing the process of lawn removal and then terraforming her yard to capture and absorb the rain draining from her roof. When prompted to assess the program, she expressed no strong opinions on her yard's reworked aesthetics ("sure, we like it") and offered measured praise for the project's role in raising her awareness of local water problems. Sure, she was busy—but she cared about such things and was happy to be helping.

When I asked about the work of maintaining the installation, however, she grew animated and expansive, recounting her ongoing struggle to keep up with it. At one point, she leaned in conspiratorially to make a confession. "I actually removed a lot of the plants, because it's just too much, they're growing out of control," she told me. "And it's crazy because when we first did it, they were all like, 'it's low maintenance, you don't have to do anything.' You do have to do stuff!" When I followed up to ask if her landscape required more effort than it had before the pilot project, when the yard had been dominated by a trim lawn, she nodded vigorously. She then explained that she'd fallen so far behind with the upkeep that she had begun to avoid spending time in her own yard, as doing so just reminded her of the work she needed to put into it.

Ariana's reflections indicate a key point of friction in visions of LA's water future relying on a retrofitted urban landscape to capture water: the necessary human labor within these arrangements does not end when the green infrastructure has been built. Long-term maintenance is necessary to sustain these installations' functionality. As such, mobilizing the work of nature requires ongoing inputs of human effort, raising key questions of who, exactly, gets tasked with these exertions and how that labor is distributed, experienced, and valued. The water capture retrofits I explore here rely primarily on a patchwork of uncompensated individual, community, and NGO efforts to maintain their functionality, an arrangement common in many other urban greening projects within and beyond LA.[60] As such, they offer revealing sites to consider the quotidian human labor that often underpins metropolitan infrastructural nature initiatives.

When it comes to large-scale infrastructures like aqueducts, reservoirs, and wastewater treatment plants, it is typically assumed that the work of maintenance falls to skilled workers employed (or contracted) by the network's owner. But distributed, nature-based infrastructures are grounded in a different spatial and material logic and demand a different knowledge base for their upkeep. As such, they are an awkward—and expensive—fit for the maintenance paradigms of institutions traditionally oriented toward another form and scale of infrastructure. The robust role envisioned for private land within some infrastructural nature programs further complicates questions of infrastructural responsibility. This combination of cost burdens, skills mismatch, and jurisdictional challenges underpins the common conclusion that public agencies are not the appropriate stewards for such installations. As in neoliberal urban governance arrangements that devolve established state functions to NGOs, private sector partners, or residents themselves, green infrastructure programs frequently assume that such entities will take on this techno-managerial work within the urban environment.

While recognizing—and questioning the fairness and efficacy of—these dynamics is key, the particularities and potential effects of this more-than-human labor also deserve attention. These are exertions that can seed new ecological imaginaries and relations. As demonstrated by recent ethnographic scholarship, cultivating urban nature's labor has the capacity to engender what anthropologist Amy Zhang has called "intimate appreciation for the vulnerability and potential of nonhuman life."[61] The agency of plants, soils, animals, fungi, bacteria, and other elements of urban ecologies is increasingly theorized as central to the development of these entanglements, offering a basis for novel urban relations to emerge even in the contexts of environmental degradation and neoliberal austerity.[62] As such, these arrangements can exceed the techno-managerial scripts commonly associated with volunteer programs of urban environmental stewardship, producing far more than just functional green infrastructures within the metropolitan fabric.[63]

Many of my environmentalist interlocutors were highly invested in the socially and environmentally generative potential of such entanglements. Some suggested that the embodied effort of pruning shrubs or pulling weeds or scooping mucky garbage from a parkway basin demands a form of attunement with one's local environment that produces new forms of ethical commitment to that landscape.[64] Others sought to incite these practice-grounded forms of ecological attention toward civic and political action, drawing residents involved with environmental care work into demanding more aggressive environmental policies and plans from city agencies. Like the land of the northeastern San Fernando Valley, the targeted energies of its residents were understood as a site of untapped potential for realizing the city's climate adaptation aspirations.

But such accounts of this labor remain incomplete if they overlook the sense of ambivalence and the burden of ongoing work that Ariana conveyed throughout our conversation about the pilot project, which echoed many others I had with Angelenos living with or near urban infrastructural nature. These are also exertions that can seed resentment and frustration with the institutions that seek to entice residents to take them on, especially within the sort of unpaid, spatially lopsided arrangements pursued in LA. Among my interlocutors, this work was sometimes described as an unfair obligation, particularly in contexts where the installations were explicitly scripted as water capture infrastructure for the city. Recognizing this dual character clarifies the complex articulations between the pursuit of urban infrastructural nature programs and environmental justice outcomes, raising thorny questions of how the benefits and burdens of new ecological work connected with such programs should be weighed.

Further elaborating these dynamics in chapter 5, I develop the term *ecosystem duties* to characterize this work, language intended to signal both a connection to the labor of nature assumed by the ecosystem services framework and the sense of obligation (and sometimes encumbrance) to a more-than-human collective that marks this work. As with the uneven distribution of environmental harms and amenities, frequently the focus of environmental justice activism and scholarship, the spatially and socially unequal allotment of such labor is imbricated in the process of sustaining radically differentiated local environments. Foregrounding the notions of fairness and value that mark these exertions allows me to track how adaptation work can both ameliorate and extend established patterns of environmental injustice within the urban landscape. Grappling with such complexity presents an opportunity to explore how relations of value and labor might be configured otherwise in arrangements of hybrid labor.

TRACING A WATERSCAPE IN FLUX

How should one conduct an ethnographic study of a water system that spans hundreds of miles and serves millions of residents? I approached the LA case with an awareness of several hurdles particular to the context. Carrying out the kind of comprehensive study that scholars of smaller water systems have conducted was not a realistic possibility. Developing an approach that would allow me to examine a range of significant nodes within the network would be essential. Further, the LADWP is notorious for its hostile stance toward researchers.[65] Gaining access to the agency's engineers and managers, undeniably central figures to the water projects I aimed to study, would thus carry the difficulties that face anthropologists studying powerful actors and institutions.[66] I designed my fieldwork methodology with an eye to addressing these challenges, attentive to the reality that my status as a relatively young white woman affiliated with an elite university shaped the conditions of possibility for establishing connections in this context.

I carried out the bulk of the research for this project over eighteen months between the summer of 2014 and the end of 2015. Building on relationships developed over the course of two summers of preliminary fieldwork, I conducted most of my participant observation research within two organizations: a local, water-focused NGO and a small company that specialized in the installation of home-scale wastewater recycling infrastructures, known as greywater systems. Working with these two organizations connected me to a range of communities, institutions, and individual residents engaged with replumbing projects across the city. My experience within both organizations proved extremely mobile, entailing flitting between company offices, water agency conference rooms, private homes, and large-scale water infrastructure sites, with many hours of conversation-filled freeway driving interspersed. My ties with the leaders of the two organizations drew me into many rooms and allowed me to build relationships with the people connected to both groups. Notably, these were not a pair of tidily discrete communities. The employees at the NGO and the company knew one another, as well as many of the same people within other water-focused agencies, NGOs, and companies.

The NGO, led by Ellen, a white woman and a longtime environmental activist, advocated for an approach to water management that would reduce LA's dependence on water sources beyond the city's borders. This phrasing is somewhat generic, because the organization pushed policies related to floodplain buyback, sidewalk repair, greywater reuse, large-scale sewage recycling plants, the parkway strips between the sidewalk and the street, and (for good measure) the kitchen sink. This range of concerns occasionally proved dizzying to absorb, but it also allowed me entry to the array of advisory groups, committees, public meetings, and site visits that united a loose community of local NGOs and water agency workers. As I trailed Ellen from her office to the Bureau of Sanitation's downtown headquarters to the oceanside offices of other environmental groups, then up to the Valley to the neighborhood pilot project, the experience of observing from within the NGO revealed the multisited nature of work for many of those focused on such issues. My work for the organization was similarly varied. Some days I helped draft and review documents, others I picked up trash from recharge installations in the retrofitted neighborhood, and often I served as a general sounding board and logistics gopher, listening to complaints and fetching lunch for visitors.

Months of observation and conversation within meetings allowed me to build rapport with Ellen's frequent collaborators, a range of NGO advocates, active stakeholders, and water agency workers. At their gatherings, someone frequently mentioned after-hours water-focused events they were attending, such as lectures at University of California, Los Angeles's (UCLA's) Hammer Museum or documentary screenings, which often led to unstructured conversations over beers and further introductions to "water people." With these friendly connections established, I pursued formal interviews with many members of these networks. By the end of my fieldwork period, I had conducted extended, semistructured, voice-recorded interviews with thirty-six water managers (a mix of agency employees and consultants, including six from the LADWP), fourteen NGO advocates, and thirteen other stakeholders.[67] Notably, in a city as racially diverse as LA, the majority of interviewees from these groups identified as white, reflecting the similarly skewed demographics I observed among these communities of practice throughout my time in the field. In addition, near the end of my fieldwork period, the NGO arranged for me to carry out interviews with twelve of its neighborhood pilot project's twenty-four participating households, most of whom I had met several times before through work for the organization. Consistent with the makeup of the project's northeastern Valley neighborhood, most of this group identified as nonwhite, with a Latinx plurality. As such, I was able to gather data from both the communities managing and working to change the water-scape, as well as those living with some of those changes (albeit more so with the former).

My work for the greywater company took a slightly different form. During preliminary fieldwork in 2013, I completed a weeklong installation training course led by a greywater-focused NGO. As a result, I had a rudimentary grasp on greywater construction techniques and was able to help work crews dig trenches and lay pipes on job sites. I spent roughly half my time with the company participating in installations and workshops and the rest accompanying Hank, the company's principal, on site consultations. A warm, quirky fortysomething white man who had worked as an architect before shifting to greywater installation, Hank also welcomed me along for many casual lunches and errands on the road between these meetings. Over the course of my fieldwork, I participated in more than thirty of these preliminary client visits, which were widely distributed across LA County. I complemented this participant-observation work with formal, semistructured interviews about home water practices and local environmental politics with twenty Angelenos who live with home greywater systems, and informal conversations with several other greywater adoptees. I pursued interviews with individuals I met at greywater talks and installation events and through other interlocutors, as well as former clients of Hank's. The stories from this community of practice take center stage in chapter 2, in which I explore the forms of domestic-scale waterscape retrofitting and work that staffers at the LADWP and other local water agencies frequently disparaged, a contrast that helps to clarify the spatial and class dimensions of the green infrastructure labor explored in the chapters focused on recharge projects.

I complemented my fieldwork with substantial archival research, reviewing materials from the City of LA housed at UCLA, from LA County housed at the Department of Public Works building in Alhambra, and from the private archive of a longtime local water advocate, which he donated to my project. In addition, I used ProQuest's database of historical *LA Times* coverage to review hundreds of articles on stormwater and wastewater. I foreground these materials in chapter 3, which traces back to the late nineteenth century the notion of managing land in the northeastern San Fernando Valley to capture stormwater. As this timeline attests, while the terminology may be new, the notion of infrastructural nature is not unique to the twenty-first century—and considering the recurrence of such concepts can help to situate analyses of climate adaptation projects, so frequently wrapped in the language of innovation and novelty.

PLAN OF THE WORK

The book's narrative unfolds through two sections, each aligned with a different genre of urban water. Part I, "Reuse," examines attempts to reroute the city's wastewater. Foregrounding the hydrosocial imaginaries— of California's past and future environment, the urban water system, and the agencies and residents connected with it—that guide this work, these chapters elaborate the grid-mediated relations that state-led replumbing efforts seek to sustain.

Sewage treatment plants and public water managers take center stage in chapter 1, which tracks a prominent element of LA's water supply augmentation plans: municipal-scale wastewater recycling projects. Among this community of practice, sewage flows are widely understood as a desirable future water source for LA. Sketching the contours of this collective affection, I show how the ideal of water reliability, which these interlocutors understand as the uninterrupted, universal provision of unlimited water on-demand to paying customers, mediates water managers' approach to climate adaptation in the region. These workers view wastewater as a steadily produced, locally controllable, infrastructurally contained source of new water for the city—perfect for sustaining their ideal of a water grid that remains reliable under conditions of heightened hydrological variability. Tracking two genres of public critique of the city's reuse trajectory—on the grounds of purity (too gross) and inadequacy (too slow and insufficiently ambitious)—I also explore how water managers seek to manage understandings of their plans that contradict or exceed their hopeful narratives. Detailing these perspectives helps to situate the allure of this approach to replumbing the city's water grid for public agency actors.

Chapter 2 considers the same waters from a different vantage, entering the world of home greywater system installers, advocates, and adoptees. Examining the distributed water reuse arrangements that public water managers rarely encourage, the chapter highlights the subtle forms of water grid control that these agencies seek to retain while pursuing new, local water supplies. Greywater systems entail a disruption of existing home-scale water infrastructure by rerouting wastewater from the sewer system to the yard. In the process, these systems can draw critical attention to established networks of water management by unsettling configurations long taken for granted and revealing the role that institutions and infrastructures of water management script for LA water users. Further, their proponents work to mobilize them as sites for the cultivation of oppositional engagements with public water agencies, contesting their policies, expertise, and legitimacy. Long-term water agency resistance to greywater adoption underlines the fact that the systems represent a mode of living with and understanding urban water that challenges the city's established management arrangements. Put differently, home greywater systems prompt residents to rethink, revise, and sometimes actively contest the roles that public agencies have within the city's water grid.

Part II, "Recharge," focuses on the rain that falls within the city, elaborating a range of aspirations and projects to direct it to the aquifers of the far northeastern corner of LA. Centered on the work of producing urban infrastructural nature to realize the city's recharge ambitions, this section highlights the hybrid labor and processes of valuation constitutive of that approach to supply augmentation.

Drawing on historical newspaper articles and other archival materials, chapter 3 traces the idea of managing the northeastern San Fernando Valley as a landscape dedicated to stormwater capture and storage back to the nineteenth century. In such visions, this section of urban fabric emerges as key terrain within the project of boosting local water supplies, due to the porosity of the local soils and the presence of the aquifer. In recent decades, LA-based environmental advocates and water managers have increasingly understood this peripheral slice of the city as capable of doing crucial water-provision work for the city. Examining the ascendance of this paradigm among LA-based environmental advocates and water managers in recent decades, I show how advocates increasingly urge a redesign of sections of the metropolitan terrain that allows "nature," in the form of carefully configured arrangements of gravel, mulch, soils, and plants, to provide selected ecosystem services—without the financial burden of acquiring new parcels of public land dedicated to that purpose. The process entails approaching new spaces, substances, and actors as embedded within the city's infrastructure for water provision—increasingly understood as its infrastructure for buffering the metropolis from ecological shocks. Situating these visions of carefully cultivated stormwater terrain, the chapter also elaborates the enduring function of this section of the city, detailing the area's constitutive role as a racialized, pollutable, underdeveloped internal periphery within the metropolis, raising complex questions about the equity and environmental justice implications of such interventions.

Interpreting a landscape as a potential buffer zone is one thing; realizing that vision via green infrastructure development is another task entirely. As this formulation suggests, enrolling the urban landscape into the work of corralling water for the city is an undertaking rife with frictions related to the flows of both water and capital. Tracking how a range of water agency and NGO actors pursued this work of spatial transformation in the northeastern Valley during the drought years of 2014 and 2015, chapter 4 explores the incomplete process of rendering infrastructural nature investable by public agencies. Developing these distributed installations required advocates to do sustained translational work, laboring (sometimes in collaboration with the material stuff of the installations) to fix particular understandings of green infrastructure and its efficacy and value in city plans and policies, as well as in the public imagination. I show how the perpetually unsettled relationship between urban runoff, infrastructural nature, and monetary valuation constrained the pace of developing these facilities—which most of my interlocutors considered agonizingly slow.

Chapter 5 focuses on the distribution of long-term forms of waterscape work emerging through replumbing projects, analyzing the ongoing human labor required to sustain the functioning of distributed infrastructural nature interventions. I track when and how this work becomes enrolled in networks of water management and circuits of value, arguing that the labor excluded from the gaze of capital and attendant processes of valuation is in fact constitutive of these systems of resource provision. Contrasting these state- and NGO-driven interventions to grassroots self-provisioning projects (and even the greywater systems discussed in

chapter 2), I show how these new forms of work help reinscribe and reproduce the city's long-standing patterns of socio-environmental inequality,
creating new burdens of free labor for residents of the northeastern Valley.

In the book's brief epilogue, I reflect on how LA's replumbing efforts
have evolved in the years since my fieldwork concluded and offer a short
meditation on how a labor lens might productively reorient accounts
of environmental management and justice under conditions of climate
change. Realizing just climate futures, I suggest, demands grappling with
the uneven, processual nature of environmental harm, remediation, and
adaptation.

PART I Reuse

1 Public Agency Work

CENTRALIZED SEWAGE RECYCLING
AND THE ALLURE OF RELIABILITY

Sipping from my plastic cup of treated sewer water, I had to admit that the guide was right: it tasted a bit flat. The advanced wastewater processing plant that our group had been touring, he explained, would eventually address the flavor issue by adding minerals to the effluent before transporting it to a nearby groundwater basin for storage. But we were ending our visit to the facility with samples of this version to experience the uncommon taste of water in an extraordinarily pure form, fresh from a lengthy cleansing regimen that featured reverse osmosis filters and disinfection via ultraviolet light.

Sweating in the summer sun, our tour group downed the water samples without much comment. The nine of us, all connected to the City of LA's Recycled Water Advisory Group (RWAG), had made the trek to Orange County's Groundwater Replenishment System that Wednesday morning to observe the process of transforming wastewater effluent into drinking water up close.[1] For most of the attendees, however, the concept of consuming treated sewage was a very familiar one. A mix of staffers from the City of LA's Bureau of Sanitation and the LADWP, advocates from local environmental NGOs, social science graduate students, and a couple of unaffiliated but interested residents, the assembled group understood that this visit was meant to serve as a preview of their city's water supply future. If LA followed the course of action outlined in its planning documents, similarly cleansed sewage would be flowing from the city's taps in the coming years.

For the LADWP staffers who had coordinated the visit, this was a preview staged with a clear purpose. Developing municipal-scale advanced wastewater treatment infrastructure would likely cost billions of dollars, so they were doing long-term public outreach work in hopes of ensuring that Angelenos would welcome this new water resource into their homes without complaint. The RWAG, established in 2009, was a centerpiece of those efforts. With an official membership of more than seventy during my fieldwork period, the group had been conceived by staffers within and consultants for the LADWP as a forum to develop sustained relationships with stakeholders engaged—both critically and supportively—with the issue. During my main stretch of fieldwork, the agency convened a handful of presentations and field trips to keep members up to date on the city's wastewater reuse plans and to address concerns related to those programs. After exchanging a few emails with LADWP staffers and taking part in an hour-long orientation session, I became official RWAG observer near the end of 2014. I proceeded to attend nearly all of the group's 2015 events, in the interest of understanding the nature of this outreach work.

Edgar, an LADWP engineer who worked with the RWAG throughout this period, acknowledged that the whole operation was grounded in the agency's fear of a resistant public protesting a sewage recycling facility all the way to closure. Several such reuse projects across Southern California, including one within LA's borders, had met this fate in the 1990s and early 2000s. Most of the criticism during that period was framed in terms of health and safety, expressed on the register of revulsion and fear at the prospect of consuming "toilet-to-tap" water.[2] But Edgar had noticed a different sort of stakeholder objection during his time working with the RWAG: many members affiliated with environmental NGOs complained that the city's plans for large-scale wastewater recycling were insufficiently ambitious or simply too slow, incommensurate with the problems they understood to threaten (or be produced by) LA's existing water system. These attendees expressed no purity concerns but plenty of disgust at the city's continued reliance on water piped into LA from distant landscapes.

In the midst of preparations to become an official LA Aqueduct tour guide for his agency that year, Edgar was somewhat sympathetic to this position. He spent time along the pipeline as part of his guide training and observed firsthand how desiccated the combination of city extraction and drought had left its source watersheds. From that vantage, he acknowledged, LA's water provision network, despite its sprawling nature, looked surprisingly fragile. "To think that's a guaranteed source—that's naïve," he once told me, gesturing toward a map of the aqueduct on his office wall. "If we don't realize that as a group, then we're going to put more stress on these already expensive and unreliable sources of water. The goal of the department is to make sure that water reliability is protected, and to do that, we need a diversified water portfolio."

The guides leading our RWAG group around the Groundwater Replenishment System echoed this point, presenting treated sewage as a desirable water source in terms of both safety and steadiness. In his opening remarks to the group, a staffer at the plant flashed a PowerPoint slide titled "Why Do We Need the GWRS?" on the screen and then offered a rundown of the reasons: we were in (yet another) protracted drought, Southern California's imported water sources were overextended, and environmental regulations were keeping some of that dwindling source water from flowing to the region. And despite all that, people here continued to rinse dishes and flush toilets, so wastewater kept sliding into centralized treatment plants. Drawing this contrast between the consistency of urban sewage flows and the variability of the region's rainfall, the speaker was making an argument that his facility's effluent should be understood as suitable in biological, chemical, spatial, and temporal terms—as a water source not only pure and safe but also local and (perhaps most importantly) reliable.

As we will see, not everyone in LA shares the view that water system steadiness, which water managers typically gloss as "water reliability," should be the primary index of a water source's desirability. This chapter considers questions about how water agencies prioritize their efforts in the face of climate change, how the LA public reacts to the idea of recycling sewer water, and how water managers seek to influence those reactions. It looks critically at how the reliability ideal mediates these agencies' approach to climate adaptation, circumscribing the water system configurations they consider possible.

While popular discussions of water systems tend to foreground questions of volume, the temporal patterns associated with the resource's presence also orient the work of its management.[3] As the history of LA's water network attests, fluctuations within local precipitation regimes—in particular, unpredictable swings between drought and deluge—can shape programs of infrastructure development. And so too can the rhythms quietly embedded in expectations of water availability within an urban grid. In the case of piped water supplies, complex infrastructural and institutional networks mediate patterns of resource availability, and persistent presumptions of the steady nature of a "modern" metropolis condition them.[4] Grounded in such urban imaginaries, in LA most residents assume that their city's water system can and will provide an uninterrupted, on-demand supply of safe, potable water for all paying customers connected to the network. Likewise, local water managers approach water reliability as a bedrock goal.

Elaborating the subtler contours of the urban imaginary of resource reliability and its effects on public water planning processes situates the project of replumbing the city in its full temporal context, surfacing understandings of infrastructural pasts, presents, and futures that structure the efforts. Cognizant of the material benefits that a steady water provision system brings to urban lives and economies (key among them being resident access to water with minimal daily labor), I use this rhythmic lens to examine the range of temporalities embedded within water management.[5] As anthropologists have shown, processes of resource extraction, consumption, and disposal are deeply imbricated with people's understandings of time, particularly their conceptions of futurity.[6] Accordingly, the looming specter of city water systems running dry has long figured prominently in politics and discourse about urban water management, a tendency that has only accelerated under climate change.[7] Locating such crisis imaginaries and actions in relation to their assumed opposite (that is, consistent, unpunctuated water deliveries) clarifies key temporal and spatial logics underpinning water planning efforts that seek to manage an increasingly unsteady hydrological setting. I trace the ways that an orientation that prizes a steady grid guides water managers' assessments of existing provision infrastructures and potential new water resources alike.

Figure 4. Effluent-settling ponds at a city of LA wastewater treatment plant. Photo by author.

This focus is especially prominent in the LA context. Water reliability looms large in technocratic discourse about and plans for the city's water future, conditioning public agencies' approach to resource development. In particular, the desire to sustain the network's reliability featured frequently in my water manager interlocutors' glowing accounts of large-scale sewage recycling facilities and the new drinking water they could provide for LA. Anticipating heightened precipitation variability across the city's sprawling waterscape, these workers, in informal conversations and official agency plans alike, presented wastewater as a readily available, consistently produced, locally controllable source of new water for the city and thus appropriate for sustaining their steady systemic ideal. Such discussions of and plans for recycled water underline the centrality of spatiotemporal rhythms to technocratic efforts to buffer LA's grid from heightened stressors.

The appeal of these reliable flows, however, is not considered so obvious beyond these communities of practice. Eruptions of public displeasure at the notion of drinking recycled sewage and at LA's specific plans for it have complicated the city's efforts to realize its sewers' water resource potential for decades. Examining two types of resistance to the city's official reuse trajectory—on the grounds of questionable purity and insufficient urgency—reveals the substantive work that goes into managing understandings of agency plans that contradict or exceed hopeful technocratic narratives of a reliable, climate-adapted urban water grid in the making. Such critiques also highlight the circumscribed scope of water system transformation that the city is pursuing. As such, elaborating these frictions helps to situate the allure of centralized wastewater recycling for established institutions of water management and to clarify the forms of stasis that this approach to provision quietly buttresses within the network.

THE RELIABILITY IDEAL AND THE WASTEWATER FIX

Edgar's characterization of the LADWP's guiding goal—protecting water reliability for LA—is written directly into the agency's mission statement: "The Los Angeles Department of Water and Power exists to support the growth and vitality of the City of Los Angeles, its residents, business and the communities we serve, providing safe, reliable and cost-effective water and power in a customer-focused and environmentally responsible manner." The aim of maintaining such systemic dependability is also the organizing principle of the state-mandated Urban Water Management Plans (UWMPs) that the agency develops every five years. Within these documents, the agency seems to use the phrases "water reliability," "water supply reliability," and "water service reliability" interchangeably. But in all cases within the document, the usage appears aligned with the relatively terse definition of "water service reliability" articulated in California's UWMP guidance document: "Water service reliability reflects the supplier's ability to meet the water needs of its customers, including end-use customers and Retail Suppliers, with water supplies under varying conditions."[8] Outlining the agency's plan for maintaining this reliability in the decades to come, the LADWP's UWMP details the city's existing and potential potable water sources, along with provisional plans for tapping into the latter should the circumstances require. Similar framings of water reliability are articulated in the city's Sustainability pLAn (2015), Stormwater Capture Master Plan (2016), One Water LA 2040 Plan (2018), and Green New Deal (2019).

In these plans and many social science accounts of California water management, reliability is frequently framed as a straightforward, volumetric condition: having enough water within the system to meet demand at a given moment, even under conditions of hydrological stress.[9] But the less-mentioned spatiotemporal dimensions of the category are also important. A reliable water network is a steady, constant system, one capable of providing an unpunctuated flow of water deliveries to every user connected to the grid within the service area. While overall water volume is obviously crucial to such an arrangement, so too is the constancy of provision it assumes. Ethnographer Lisa Björkman's articulation of the engineering ideal of network flow, an orientation that envisions "the movement of some substance through space (over time) while attending to the present-time work of implementing plans that might enable such flows (through space) at some imagined future time," is helpful in highlighting these dynamics.[10] Maintaining reliability is conceived as a project of ensuring that unrestricted water is always available on demand to customers, constrained only if a user fails to pay their bill.

Understood in these terms, the notion of water reliability guiding the work of this community of practice is comfortably aligned with that of what geographers Stephen Graham and Simon Marvin term the *modern infrastructural ideal* of universal, centralized urban water provision and disposal for housed, paying customers, an arrangement described elsewhere in the literature as the "sanitary" or "bacteriological" city.[11] Grounded in what urban political ecologist Maria Kaika has termed "Modernity's Promethean Project" of disciplining and taming nonhuman nature, this approach to rationalizing metropolitan space through water management assumes a high degree of state control within the urban landscape.[12] Compared to other arrangements, it also demands minimal waterscape labor from residents connected to the water grid, as they neither need to periodically transport nor temporarily store water for domestic use. Water supply procurement and distribution becomes the purview of a formal, often public sector, and sometimes unionized workforce, with much of the LADWP and MWD fulfilling all three descriptors.[13]

My water manager interlocutors treated the smooth constancy of water flows from LA's taps as the signature achievement of their region's sprawling water provision network and as the state of being that all desired to sustain. Many also took pains to remind me that, if I cared to examine the region's history, I would encounter a remarkable record of success in doing so. Southern California water agencies' consistency in maintaining reliable water service despite hydrological variability was a point of pride that surfaced frequently in our discussions of climate adaptation and water planning. So was the established pattern of steady water provision under recurrent ecological stress. In my conversations with Nelson, a midcareer staffer at MWD, he loved to draw our focus back in time. This tendency might initially seem surprising for someone whose day-to-day work during that period focused on his agency's integrated resources plan (IRP) for long-term water source development, a fundamentally future-focused project. But as he clearly conveyed when we met in his office at the MWD's downtown LA headquarters, a celebratory account of the agency's approach to past droughts guided those future-oriented efforts.

Nelson had joined Metropolitan in the mid-1990s, shortly after the end of the state's 1987–1992 drought and just as the agency was developing its first-ever IRP. Reflecting on that work and all the IRPs that had followed, he emphasized his agency's sustained success in realizing the reliability ideal through enormous regional population growth and recurrent multi-year dry spells. The IRP's short-term response to the drought in the early 1990s, a period of protracted aridity that nearly triggered the first-ever across-the-board cuts in deliveries to its member agencies, was a massive water storage project: the development of the Diamond Valley Reservoir. This manmade lake can hold eight hundred thousand acre-feet of imported water—substantially more than the total volume of water consumed annually within the city of LA in recent years—in reserve to buffer the region against dry stretches. Such liquid stockpiles, Nelson told me, help preserve the tempo of water deliveries to end-use customers in years when the aqueducts deliver little water, eventualities that most expect to see more of as the climate changes.

In this framing, an altered climate figures not as an existential threat but as an intensifier of well-established patterns of hydrological inconstancy, conditions that water managers believe themselves to have been dealing with for decades.[14] For Nelson, this understanding led to considerable skepticism about the idea that climate change should spur a major departure from past practices of water planning and development within the region. "I'm unwilling to jump on the panic train," he told me, before launching us into a discussion of how the current drought had ratcheted up public pressure to accelerate the agency's carefully vetted plans for developing new water sources. Metropolitan, he explained, had been making steady progress on a long-term scheme for establishing wastewater as a new water source to sell to its customers across the region for years. To him, the record-shattering dry spell crisping lawns in the mid-2010s simply reaffirmed the wisdom of that path, rather than suggesting a need to redouble or accelerate such efforts.

In our conversations, water managers like Nelson frequently described as disastrous and unfamiliar the idea of shifting to a system of rationed or punctuated water deliveries within Southern California's cities, and said they were working hard to avoid that situation. They often articulated fears it would emerge by referring to the conditions of "third world" cities and towns, a formulation that overlooked the fact that the water supplies of several marginalized communities within California were running dry as we spoke.[15] Typically, they discussed such undesired scenarios of scarcity at the scale of the individual water consumer and in terms of the quotidian experience of the water provision network. As Nelson put it: "My definition is about the end use customer not having an interrupted water supply, that doesn't have fines and penalties—no shortage rationing. Can the customer go about his day-to-day life without worrying about the shutoffs at all? That's the goal of reliability." Articulated in these terms, systemic reliability clearly overlaps with the much-theorized condition of resilience, signaling the system's capacity to continue functioning without disruption in the face of exogenous shocks.[16]

Such characterizations assume a common experience of LA's current water grid among customers, one marked by steady, secure resource availability. While acknowledging that most LA residents do enjoy such conditions, there are persistent exceptions to this generalization within the space of the city. Southern California is a high cost of living, rent-burdened region, where thousands of households struggle to pay their water bills and are thus forced to navigate service shutoffs.[17] Recent research on "plumbing poverty" demonstrates that over ten thousand housed residents lack access to clean, safe water within the city of LA.[18] Further, while almost all apartments and houses of LA may be connected to the water grid, a sustained housing crisis means that tens of thousands of Angelenos lack access to shelter—and thus to steady potable water flows.[19] These conditions conform with a growing body of work from scholars of water insecurity in the global North, who have shown that the modern infrastructural ideal of safe, adequate, universally accessible piped water remains elusive, even in wealthy and water-rich contexts.[20]

Identifying the persistence of household-level water insecurity within LA helps to clarify the scale of the water reliability imaginary that guides my interlocutors' work. While these water managers' understanding of the concept is rooted in a desire to sustain uninterrupted deliveries for all residents, their efforts are primarily oriented toward the particularities of citywide water supply capacity, rather than the lived experience of home-scale water (in)accessibility. That is, the systemic dependability that they seek to preserve operates on a scale distinct from the household-level reliability that eludes thousands of marginalized LA residents. This approach was clear in our conversations and reflected in plans like the UWMP, which focuses primarily on how to obtain an adequate volume of water to consistently meet projected demands within the city, effectively equating the presence of supply with universal ability to access those flows.[21] Such a perspective means that the perceived reliability of a water source—established or potential—guides these water managers' assessments of its desirability. And as the commentary on our tour of the Groundwater Replenishment System suggests, this community of practice understands recycled wastewater as an ideal addition to the region's supply mix, due largely to its perceived steadiness. These interlocutors tended to treat safety as a settled issue and to focus instead on the issue of reliability.

Carl, a prominent consultant in the field, was particularly forceful in this sort of redirection. "We have the technology to convert any quality of water to any other quality of water, it just depends on what you're willing to pay," he told me with a firm nod one late summer morning. Though we'd crossed paths only once before, it wasn't the first time I'd heard him make such a claim. A couple weeks earlier I had attended a public panel on engineering solutions to drought, hosted by a media company at an open-air downtown food hall. Upon arrival I was quickly intercepted by Hal. An avuncular white retiree and RWAG regular, he observed happily that we'd be hearing from two different water reuse experts that evening.

Carl was half of that pair. He spent most of his speaking time singing the praises of recycled wastewater, describing it (as many of his colleagues did) as an essential element of a "diverse portfolio" of water sources for the region. When the moderator raised the specter of health concerns, he responded with the "convert any quality of water to any other quality of water" line, well-practiced at dismissing such worries with appeals to technological competence.

Over beers with Hal and me after the panel and again in a follow-up interview, Carl made it clear that he thought we should be discussing water reuse in terms of its role in sustaining reliability, not wasting our breath on the topic of safety. When I pressed for his working definition of water reliability, he replied that the best approach to the concept is to focus on the relative dependability of the source water. If I understood this point and the purpose of urban water management—to keep liquid sloshing steadily through the distribution network at all times—then I would be able to grasp what makes a good twenty-first-century water source. In his view, this perspective would help me appreciate why he and his colleagues see sewage as such a crucial element of Southern California's water future: it's far steadier than the region's rain and snow, especially these days. Expensive to develop, sure. But eminently doable, and worth it, given the dependability of the source. Seawater, of course, would be even more reliable—but given the higher cost and carbon emissions associated with purifying it, the case for wastewater as the better immediate-term option was clear.[22]

A water reuse maximalist, Carl unapologetically advocated for an approach to wastewater recycling called *direct potable reuse*. In such systems, water can be transferred directly from an advanced treatment facility into a jurisdiction's drinking water distribution network. *Indirect potable reuse* projects, in contrast, incorporate an intermediate period of environmental storage into their process. At the Groundwater Replenishment System, for instance, the cleansed effluent is spread over porous soils, directing the liquid into a subterranean aquifer for storage before it is pumped up, treated, and piped to water users. The same punctuated spatiotemporal rhythm of flows marks all existing wastewater-to-drinking-water facilities in California, an arrangement that regulatory requirements help drive.[23] The state only approved formal requirements for direct potable reuse projects in late 2023, due to safety concerns associated with the effluent cleaning processes. As a result, at the time of this writing most plans for wastewater recycling projects in the state incorporate intermediate environmental storage in some way.

Drawing out this technical detail helps to clarify a rhythmic point: in this context, storage arrangements significantly condition continuous water flows within the grid. On one level, this is nothing new, as reservoir and aquifer storage have long shaped (and been shaped by) processes and patterns of urbanization in the region.[24] And while much of the laudatory talk about wastewater emphasizes the constancy of those flows in a manner that suggests a direct temporal connection between effluent treatment and drinking water distribution, the existing plans for that substance are oriented instead toward bolstering subterranean water stockpiles. For instance, the LADWP describes the purpose of one recently proposed water reuse project as building "a resilient storage supply in local groundwater basins."[25] As chapter 3 elaborates in detail, the San Fernando Groundwater Basin is a key target for such water stockpiles, which water managers intend to expand through both storm and wastewater recharge.[26] Notably, despite the popularity of the analogy, groundwater basins are not actually storage tanks, and the waters within them move in complex, not entirely understood ways.[27] For now, I simply seek to flag these constitutive moments of water's (relative) stillness within this network of flows and the key role of local aquifer storage in the region's arrangements of water provision more generally. Because while rhythm is clearly salient to the wastewater imaginaries elaborated here, the constitutive tempo is somewhat more complicated than many characterizations suggest.

Though annoyed at what he saw as state regulators' slow progress toward allowing direct potable reuse, Carl was sanguine about the prospect of wastewater recycling shoring up LA's supply in the coming years. "I don't think that there will be a worst-case scenario," he told me near the end of our interview. "The history here: we solve the problems, do it in intelligent ways based on the available technologies." As he acknowledged, however, when it came to recycling sewage, this statement has been truer in some Southern California jurisdictions than others. Over the years, public outcry has stymied several marquee reuse projects, a pattern of failure that has fed the anxieties and structured the workdays of many public agency staffers.

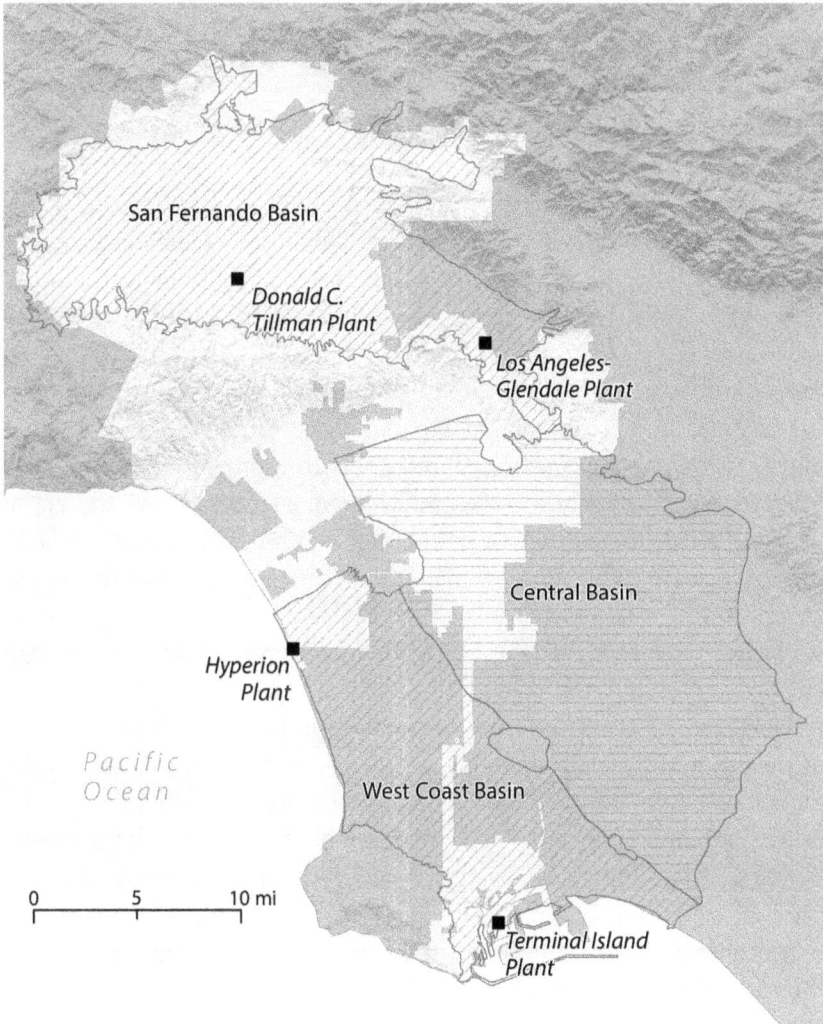

Map 3. City of LA's four wastewater treatment plants, depicted in relation to the three local groundwater basins under consideration for effluent storage arrangements. Map by Nick O'Gara.

WASTEWATER AS A (POTENTIALLY) RELIABLE SOURCE

In Frank Herbert's classic sci-fi novel *Dune* (1965), characters don full-body outfits known as stillsuits when they venture into the interior of the desert planet Arrakis. The suits enable a critical bit of resource circularity, absorbing and transforming the wearer's bodily wastes into drinkable water. Though it was years before Denis Villeneuve's blockbuster movie adaptations reintroduced the text to a mass audience, the novel came up repeatedly during my time among LA-area water managers. Those who mentioned it typically compared the characters' stillsuits with the region's wastewater recycling plants, carefully noting the scalar adjustment from the waste of one suited individual to that of the entire city. The commonality is that waste is transformed into a resource, providing a steady flow of water no matter the weather. Such literary allusions give the act of drinking purified piss-water a sci-fi sheen, reframing it from a stomach-turning moment of intimacy with an urban waste stream into an embodied encounter with an innovative, futuristic approach to water provision. Recycling wastewater would thus enable LA to remain in its long, high modernist present of reliable supply.

However, for years there has been a key problem with such accounts of the city's water trajectory: LA, unlike many of its neighboring jurisdictions, had already tried and failed to incorporate treated wastewater into its potable supply mix.[28] The metropolis's stumbling path from identifying to exploiting a new resource aligns with ethnographic accounts of resource extraction in other industries and other places.[29] The local history of sewage-to-drinking water facilities within LA reiterates the contingency of such processes and suggests that an unruly public has proven skeptical of the water managers' vision of this futuristic but reliable version of the city's water grid.

While the possibility of a municipal-scale sewage-to-drinking-water project surfaced and resurfaced in the city throughout the twentieth century, the idea finally began to gather institutional motion during the record-breaking 1987–1992 drought. In 1989 LA created a temporary Office of Water Reclamation and tasked the agency's staff with studying the feasibility of reusing wastewater from the city's sewage treatment facilities. By the mid-1990s, so-called water reclamation work had moved into the portfolios of two permanent city agencies, the Bureau of Sanitation and the LADWP. Yet at the time of my fieldwork, while LA used treated effluent for limited irrigation and industrial purposes, cleansed wastewater was not yet part of the city's potable supply. The major roadblock to progress toward this goal came in 2000, when public complaints led to the almost-immediate shuttering of the East Valley Water Reclamation Project. Connected to the city's Donald C. Tillman Treatment Plant, the infrastructural complex was designed to pump treated sewage to the San Fernando Valley's stormwater spreading grounds for percolation into the groundwater basin beneath them.

This failure weighed heavily on the minds of my LADWP interlocutors. During a question-and-answer period at one RWAG gathering, Morgan, a senior staffer at the agency, spoke about the project's demise with palpable feeling. Nearly a hundred of us had assembled in an auditorium in a nondescript city building in the San Fernando Valley, and the room quieted as he recounted the bruising process. Back in the day, Morgan had been involved in developing the facility, work that carried a particular sense of urgency after the drought of the early 1990s and the 1994 state ruling that limited LA's take from the Mono Basin. The LADWP had modeled the project on a similar wastewater-treatment plant-to-pipeline-to-spreading basin-to-groundwater-basin assemblage that the LA County Sanitation Districts had operated about twenty miles beyond city borders since the 1960s. Given that plant's lengthy track record, few city staffers anticipated a robust public outcry against a doppelgänger plant.

This miscalculation continued to haunt Morgan. The fact that the East Valley facility was allowed to operate for just a single day was "a real heartbreak for me," he told the group. He then spent several minutes detailing how technological advancements in the intervening years mean that water reuse projects are now even safer than ever, before concluding with a few words of warning: "No project is worth anything if it's not going to be accepted by the public. There's another element, it's not technical, it's emotional."

Public acceptance, in such accounts, emerges as a powerful threat to the project of tapping into this new supply source and shoring up water reliability for the city. The tenuousness of residents' approval of new projects was frequently attributed to an instinctual revulsion at the idea of reencountering water that had previously contained human waste. Such an understanding was invoked often in settings where water recycling professionals gathered to mingle and share stories. Sitting through two days of keynotes and panels at the WateReuse California conference, hosted in a downtown LA hotel in 2015, I watched my local water manager interlocutors and their colleagues from across the state recount community outreach challenges. Explaining public skepticism, many mentioned a notion known as the "yuck factor," understood as a widely shared aversion to sewage that "renders recycled water unpalatable from the public's point of view."[30] Wastewater, in such framings, surfaces as a potent, lively substance—much like in recent ethnographies of sewage management work in African and Asian cities.[31] But in contrast to such analyses, with their emphasis on waste materials' complex, world-making effects and political imbrications, the speakers at the conference tended to focus on a single question: How can we ensure that people overcome this supposedly intrinsic response to such a substance?

Most speakers framed the work of taming revulsion as a project of educating away a reaction that they considered understandable but fundamentally flawed, given the quality of the cleaning technology involved. Such education efforts often involved cultivating the kind of embodied encounters with treatment infrastructure and just-cleansed wastewater that I experienced at the Groundwater Replenishment System—which is to say, tours of established effluent-reclamation facilities. But simpler tactics were also cited. Presenters repeatedly referred to preemptive outreach to community groups and media outlets to cultivate trusting individual relationships with staff at the water agency pursuing reuse. They also often mentioned outreach materials that emphasized the established nature of wastewater recycling facilities in other cities. And when in doubt, as one speaker said, "It never hurts to talk about the space station"—and, more generally, to refer to the fact that astronauts consume highly treated urine as a water source.

Edgar, who sat through several of the WateReuse sessions alongside me, was vigilant about criticism grounded in revulsion or hostility at the prospect of drinking treated wastewater. While public declarations of such fears had ebbed since 2000, they continued to bubble up throughout my time in LA. One memorable eruption occurred in a City Council election during the winter of 2015. Alerted by a RWAG email, one weekday morning I headed to a lightly attended mayoral press conference celebrating the expansion of the city's nonpotable recycled water network to a public golf course. Milling around on the grass, paper plates of complimentary fruit salad and muffins in hand, Hal and I made our usual recycled water small talk as we waited for the speeches to start. But before the pomp and circumstance could get underway, Edgar sidled over to tell us about a concerning development. An unknown candidate in a South LA City Council district race had decided to make resistance to wastewater recycling a key plank of his platform. And he was dedicated enough to the message to erect a banner emblazoned with the words "No Toilet-to-Tap" across a major neighborhood thoroughfare. Half-watching as the mayor's team fiddled with the makeshift lectern and sound system, Edgar explained the agency's current approach to dealing with the critic. For the time being, they would simply monitor his statements and media coverage—he didn't appear to be gaining much support, so there was no sense in raising his profile by responding publicly just yet. But if the candidate seemed to catch fire, they would reach out to members of the RWAG to speak as advocates for the city's recycled water program and its expansion.

Interest piqued, a few weeks later I attended a Sunday afternoon pre-election forum for the district and listened to the candidate regale the audience with stories of LADWP's dishonesty and the dangers of drinking inadequately treated sewer water. My fifty or so fellow attendees mostly looked puzzled or bored during these diatribes, perhaps impatient for the two leading candidates (who never engaged the recycled water question) to discuss other topics. A few weeks later, election day brought a crushing defeat for the "No Toilet-to-Tap" platform. But while that outcome had never been seriously in doubt, such periodic flareups of anti-wastewater-reuse sentiment clearly reinforced water managers' sense that fears regarding the safety and purity of their prized water source lingered within the city. In contrast to more generalized forms of distrust regarding public agencies' capacity to deliver clean water through aging urban pipes, concerns commonly managed by residents of US cities via the purchase of domestic filters, protests against effluent recycling projects provided Angelenos with a potent, prospective target for their worries. As Edgar had explained in some detail during my RWAG orientation session, the LADWP's high-touch approach to outreach on the topic is grounded in this under-standing of the situation: for all its steady rhythmic charms, within LA, wastewater's status as a resource in the making remained tenuous.

RECYCLED WATER AND THE LIMITS OF "LOCALIZATION"

While worries about such resistance persisted among water managers, by the mid-2010s a different genre of criticism was on the rise within the RWAG: the view that recycled sewage was obviously a good water source, but the LADWP was not pursuing its development at an adequate scale or pace. A case in point: during the question-and-answer period at the RWAG meeting, Morgan was responding to a query from a longtime envi-ronmental advocate named Roland encouraging the agency to dramatically *accelerate* its plans for incorporating wastewater into its drinking water supply system, in the name of better addressing the ongoing drought. "I participated in the East Valley project back in the day," Roland began. "Why are we still talking about this? Why can't we just turn the East Valley Project back on?"

The next two questioners reiterated his sense of urgency, badgering the assembled city staffers to move more quickly toward implementing indi-rect potable reuse within the city to help it get off imported supplies, bat-ting aside Edgar's earnest interjections about the need for careful outreach to ensure broad public acceptance. After the formal program concluded, Roland approached a gaggle of fellow environmentalist stakeholders to share his plan for circumventing the LADWP's dallying. He had an in with a termed-out city councilmember who had little to lose and wanted an environmentalist legacy, Roland told the group. So he was going to try leaning on the soon-to-be-ex representative to pull strings and get the plant humming the way it did for that one day back in 2000. "Things have changed in the past eighteen months!" he intoned, invoking the record-setting dry spell as justification for his scheme.

As such (ultimately unsuccessful) plots suggest, the temporal rhythms of the region's periodic dry spells are also deeply imbricated in the punc-tuated process of wastewater's resource becoming within LA. Politicians, water managers, and environmentalists alike have tactically mobilized arid stretches to justify the acceptance of and investment in a large-scale program of water reuse in the city, echoing the rationales for supply-side drought responses elsewhere.[32] Though now rarely referenced, a short but severe statewide drought between 2007 and 2009 was the explicit justification for the development of then mayor Antonio Villaraigosa's Securing LA's Water Supply (2008) plan, which revived the city's serious consideration of large-scale wastewater recycling facilities. However, while water managers and many environmentalists approached the 2012–2017 drought as a valuable window of opportunity for finally transforming sewage flows into a water supply source for LA, these actions were grounded in contrasting accounts of the appropriate spatiality of a climate-adapted urban waterscape.

Not long after the big RWAG meeting, Roland drove up to the Owens Valley for a weekend of outdoor recreation and relaxation. While there, he tracked down our mutual friend Sam Bode, who was in the middle of hiking the length of the LA Aqueduct for a documentary film she was making.[33] Shocked at how skinny Sam had grown over her weeks of trekking in the harsh terrain, Roland bought her a peanut butter chocolate milkshake and listened to her stories of navigating the lands alongside the pipeline. I played a supporting role in some of those anecdotes, having spent a week walking alongside Sam earlier in the summer. During our days together, Sam and I filled hours waiting out the midday sun under makeshift shade shelters and discussing the LADWP's tremendous footprint within remote sections of eastern California.

Observing the agency's presence within such landscapes had sparked Sam's initial interest in the aqueduct. A charismatic white woman raised in northeastern Pennsylvania, Sam had moved to LA to work as a film editor in the late 2000s after graduating from university. Not long after arriving, she spent a weekend camping near Mono Lake. There, over three hundred miles north of downtown LA, she encountered something that seemed out of place: trucks bearing the familiar city LADWP insignia. What the hell was the water department doing all the way up *here*, she wondered—a curiosity that eventually led to her sixty four day hike along the city's eponymous pipeline and the resulting film.

From our vantage during the week that I joined her project, the traces of the LADWP were hard to miss. In addition to the aqueduct itself, the surrounding swathe of the Mojave Desert was dotted with fences and

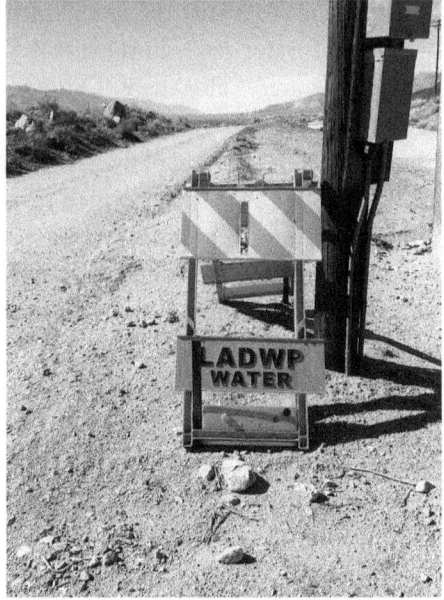

Figure 5. LADWP signage making the agency's presence known near the northern edge of the Mojave Desert. Photo by author.

placards declaring stretches of the land beneath our feet to be LADWP property. The signs got decidedly more prescriptive when we reached a reservoir near the southern tip of the Owens Valley. Our decision to ignore their exhortations to keep out resulted in a lengthy discussion with a local LADWP worker. He was friendly, explaining that the land surrounding the reservoir was technically off-limits to the public—but that in practice it was frequented by locals in search of uncrowded ATV trails and quiet camping spots, and that he wouldn't hassle us if we pitched our tent along the reservoir's banks. Nonetheless, the encounter reinforced our growing appreciation of the distant water agency's seemingly unavoidable presence in these remote rural lands.

Beyond such direct observations of the landscape and its day-to-day management, Sam had conducted interviews with ranchers, environmentalists, and representatives from Paiute tribal leadership about the destructive effects of LA's century of water extractions on the area. *The Longest Straw* (2017), her film, depicts these impacts—including biodiversity loss, particulate pollution caused by dust exposure from draining lakes, and displacement of Indigenous communities—through clips from those conversations, historic images, and extensive footage from her trek up the aqueduct. Following this narrative of hinterland degradation caused by the LADWP, the movie ends on a hopeful note. In its final minutes, Sam details a handful of in-city water conservation, wastewater recycling, and storm-water capture projects, framing them as undertakings that might enable the city to stop drawing so much water from lands at the edge of the Eastern Sierra.[34] Replumbing the city, in such a rendering, figures as a project defined by the benefits it brings to landscapes far beyond LA's borders.

This aspirational account of a reconfigured LA water provision network aligns with environmentalist narratives around the need to "relocalize" the city's water supply. Foregrounding the toll of LA's water imports on distant environments and the carbon emissions required to transport some of its faraway sources, my NGO-based interlocutors were inclined to present the city's extensive supply arrangement as not only potentially unreliable under conditions of climate change but also damaging to many communities and ecologies within its extended hydro-social territory. Drawing on many of these arguments, in 2013 UCLA went so far as to make a "100 percent local" water supply by 2050 one of its sustainability "grand challenges" for the city.[35] Imagining LA relying exclusively on in-city groundwater, captured rain, recycled wastewater, and reduced consumption to sustain its four million residents, the university's initiative was another articulation of environmentalist desire for a version of LA that would play a far smaller role in shaping distant landscapes like the Owens Valley.

This was decidedly not the vision of my water manager interlocutors. While increased local supply is enshrined as a goal in a growing range of the city's planning documents, there is little evidence to suggest that LA is in the process of relinquishing its claims to the beyond-the-city sources—and moreover, there are growing indications that the city is fighting to retain access with renewed vigor.[36] This position was reaffirmed for me time and again during my fieldwork, through interviews and public forums hosted by LADWP staff. As an agency representative described water from the LA Aqueduct system at one such gathering: "It is a very high-quality water resource for the city and we control it, so we'll continue to use it as that percentage of our water mix." The agency's most recent UWMP also reflects a commitment to continue sourcing water through that pipeline, via tables projecting that the city will be using throughout the decades ahead a larger volume of water from the system than it delivered in the drought year of 2015.[37]

Such projections demonstrate that while the perceived dependability of water delivered to LA from beyond its borders has fallen since the middle of the twentieth century, the city's water managers still assume it will play a substantial role in LA's supply mix in the years to come. When I pushed them, the LADWP employees I spoke with always acknowledged that in the long term they see both the use of LA Aqueduct water and the city's purchase of supplementary supplies from Metropolitan as inevitable. Hydrological variability might lead to radical reductions of flows from one source in a year like 2015, when the combination of low snowpack and the requirements of a dust mitigation settlement in the Owens Valley led the agency to take almost no water from the LA Aqueduct system from May through November. But the operating assumption is that in most years, both supply options will remain accessible to the city, to protect LA from widespread service interruptions even in the event of extreme water stress.

Sitting in an LADWP conference room with Noel and his colleague William discussing their work on the agency's in-progress UWMP, I had this position explained to me in particularly blunt terms. Our trio had started the conversation as a quartet, but the most senior of the three LADWP staffers had made an early exit. Noel and William, a pair of mid-level engineers at the agency, had been relatively quiet before her departure but now seemed eager to share their skeptical assessments of calls to eliminate imported supplies from LA's water mix. William, after detailing how he and his colleagues use historic water demand and delivery data to model expected future needs, framed his perspective in terms of precedent. "What supply has always come in to keep the city going?" he asked rhetorically, before answering: "It's always been MWD. So I guess, at a very high level, our reliability is pretty heavily dependent on MWD being reliable because they've always been reliable when our local supplies and the aqueduct doesn't produce what it's supposed to." Noel jumped in to reinforce the argument, using the second of the two common acronyms associated with Metropolitan. "No matter how much we like to buy MET water, [it] will always be part of our water supply portfolio," he told me. "Even in an average year scenario ... regardless of what people say: we won't be able to get off of MET."

Throughout our conversation, the pair described the city's water supply as a portfolio in need of sufficient diversity to protect users from harm. I encountered this description of water sources frequently during my fieldwork, mirroring political ecologist Joe Williams's findings in nearby San Diego during a similar period.[38] In such formulations, usually articulated at the city or county scale, relying too heavily on a single water source is presented as a dangerous, unreliable strategy and diversification as the obvious approach to managing such risks. The parallels between this framing and the way that risk is understood within investment portfolios is hard to miss, a comparison that clarifies the role that these managers assume for water sourced from in-city waste or stormwater in the years to come. Capturing, cleaning, and storing such flows as a reserve is not a project intended to eliminate use of the city's hinterland resources. Instead, the sewage cleaned and then stockpiled in the city's subterranean basins is meant to serve as an additional source, a buffer against the forms of variability threatening its other supplies.

While the critiques of imported water that politicians and environmentalists articulate might lead one to assume that LA is in the process of loosening its material ties to faraway landscapes, it is more accurate to say that the city is now establishing new resource links to terrain within the city (including its sewershed). By doing so, LA seeks to keep its consumer-facing delivery network reliable in moments when the pipelines can offer little water, continuing to provide customers with an uninterrupted flow of the resource in the face of a protracted drought, a destructive flood, an aqueduct-cracking earthquake, or even a more draconian regime of environmental regulation of its source waters. Those material connections are becoming visible and contested in new ways, in marked contrast to how they have been ideologically and discursively obscured in the past.[39] But the city's long-term effort to transform wastewater into a resource does not reject an arrangement in which the demands of the urban center shapes peripheral landscapes through extraction. As such, water managers' efforts to expand local water supply are best read as projects to add sources for the sake of reliability, not to replace current sources for the sake of independence.[40] Rather than shrink the city's hydro-social territory, water managers simply seek to intensify water production within it—a goal in tension with those of environmentalists and other critics who seek more capacious forms of systemic transformation through projects of replumbing the city.

As this chapter has shown, for my water manager interlocutors, centralized wastewater reuse projects are largely seen as a sort of infrastructural dream deferred, a key piece of the sustainable, circular (though not entirely so) urban metabolism that cities like LA should have in a climate-adapted future. The relative consistency of urban sewage flows is central to this assessment. In contrast to California's increasingly variable hydrological regime, the rhythm of the sanitary sewer network looks remarkably consistent—and thus, like an ideal source of supply for a steady urban water grid. Oriented toward preserving the reliable, unpunctuated rhythms of LA's provision system, water managers view recycling wastewater as a means to augmenting and extending established arrangements.

Expressing such positive associations with large-scale wastewater recycling facilities, these technicians were aligned with what geographer Kerri Jean Ormerod has termed a "neosanitarian" subject position on urban water reuse.[41] This techno-optimist perspective foregrounds the safety and suitability of advanced water treatment and reuse facilities, particularly in the arid US Southwest. As Ormerod notes, while such facilities unquestionably reroute urban flows, they leave major elements of the waterscape undisturbed: "The neosanitarian technological solution requires no modification of behavior for the average urbanite and minimal, if any, policy changes."[42] Geographers Valentin Meilinger and Jochen Monstadt make a similar point in a recent analysis of LA's wastewater politics, asserting that this approach serves to "reduce political questions of how water is consumed, managed, and publicly governed in Los Angeles to technical problems addressed by path-dependent engineering practices."[43] As such characterizations attest, at its core the project of centralized wastewater recycling is a technical adjustment, one undertaken with an eye to systemic stasis on several registers.

Pursuing this approach, the city's water managers advance an implicit claim about LA's twentieth-century legacy of water development: that despite any flaws, its ongoing success at maintaining water reliability for the city means that the infrastructures, institutions, and logics underpinning that steadiness are worth sustaining. This perspective was on full display in early 2019, when the city announced its intention to expand its planned water reuse facilities to tap flows from Hyperion, the city's largest wastewater treatment plant. Still following the city's water news after the end of my fieldwork, I streamed the meeting of the LADWP's Board of Commissioners that featured a presentation about the project by a staff engineer. After walking his listeners through some technical details, he concluded by calling their minds back to the region's early twentieth-century water development heyday. "To be clear, when we talk about a Mulholland moment or building a third aqueduct, this is what it looks like," he noted, before inserting the project into the city's forward trajectory: "It will make the city more sustainable, by building water storage, but also provide a hedge against the future impacts from climate change and drought." In other words, the way to keep the city's water system reliable is to think like the guy who built its original aqueduct—despite the fact that a dam William Mulholland signed off on collapsed and killed hundreds shortly after its completion in 1925.[44]

But as this chapter has also demonstrated, these are certainly not universal assessments, of the LA water system's trajectory or its plans for wastewater recycling. Orienting the work of public water management around this rhythmic ideal constrains imaginaries of potential futures for the water grid, seeding several forms of contestation and negotiation. Concerns about the safety of treated sewage water persist in some corners of the city, occasionally bubbling into prominent public view. Meanwhile, environmentalists argue that rather than functioning as a mere addition to LA's grid, recycled water should displace infusions of water from the distant landscapes that have long provisioned the city. Seeking to manage both forms of critique, water agency staff put in long hours working to convince residents that their vision for the stillsuited city of the future was safe, sensible, and reasonable—in contrast to environmentalist pipe dreams of a localized grid. Adaptation work, within this community of practice, emerges as a project of directing both infrastructure development and public sentiment. But as the next chapter shows, water agency plans coexist with alternative visions and scales of urban reuse, grounded in a divergent understanding of the proper roles for residents and the urban landscape within the city's system of water management.

2 Disrupting Water Consumption and Control at Home

On a too-hot late summer Tuesday afternoon, I entered a darkened classroom on the grounds of a botanical garden and waved to Hank, who was there to give a free public lecture about greywater recycling. Hank, a former architect, makes his living promoting and constructing home-scale water reuse systems. Shortly after I arrived that day, he took his place at the front of the room, queued up his PowerPoint presentation, and began to explain some key terminology to the handful of people gathered for his talk. The word greywater, he told us, refers to any water from laundry machines, sinks, bathtubs, and showers that is reused on-site, replacing tap water for landscape irrigation. A greywater system—like the ones installed by his company—reroutes those waters from the sewer to the garden, through a combination of pipes and sometimes pumps, he explained, clicking to a slide with an illustration of one such system.

After gliding through a description of how, exactly, these infrastructures work, Hank spent the bulk of his time detailing why he believes this technology is so advantageous for Angelenos. Displaying a slide dominated by a picture of his modest home surrounded by fruit trees, Hank told us that his affection for cutting his household's potable water consumption by reusing wastewater is rooted in the feeling that it helps make a house something other than a "malevolent presence" within an urban ecosystem. Ticking through the benefits of greywater systems, he described the pleasure he took in knowing that his son's bathwater helps replenish a depleted local water table and sustain trees that provide shade and food, rather than sliding into the sewers and eventually, uselessly "watering the Pacific Ocean."

The small audience, consisting mostly of middle-aged white women, was receptive to Hank's presentation and filled the question-and-answer period with queries about the logistics of system installation. Aware that their region was experiencing a historic drought (conditions sure to be re-peated as climate change progressed), the crowd was eager to learn more about a technology that could enable them to "do their part" to address the looming water supply crisis they kept hearing about on the news. Un-mentioned in that day's discussion, however, was a wrinkle that would likely have surprised the civic-minded attendees. Despite widespread ac-knowledgment of a precarious water supply and sustained pressure from advocates like Hank, the LADWP had long refused to incentivize or other-wise encourage home-scale greywater reuse. Such lack of interest in these conservation-minded arrangements appears even more striking in the con-text of the city's aggressive drought-time lawn removal subsidy program.[1]

A simmering conflict over the proper scale and oversight of urban reuse influences this reticence. Staff from city agencies have long sought to limit the spread of greywater systems across the city. In their view, to encour-age a perspective like Hank's, with its emphasis on using effluent to cul-tivate domestic ecologies, would undermine their own goals for the city's wastewater. By and large, workers at public agencies understand their institutions—not individual residents—as the appropriate managers of this liquid. Diverting wastewater from the grid, domestic-scale reuse sys-tems thus disrupt not only established material flows within the home but also public water managers' role in and control over the urban water grid.

Home greywater systems, like other quotidian fixtures that connect urban residents to broader infrastructural networks, serve as sites where multiple forms of techno-political relations are negotiated.[2] In LA, the roles of residents and public agencies within the city's water grid are being reimagined, reworked, and sometimes actively contested through these domestic fixtures. This chapter elaborates the moral ecologies and

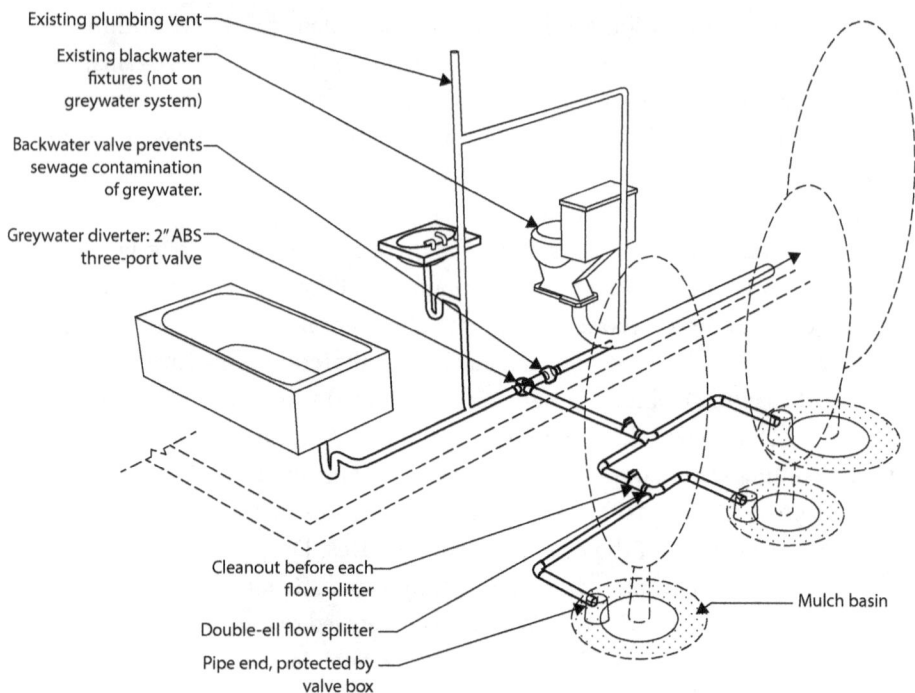

Existing plumbing vent

Existing blackwater
fixtures (not on
greywater system)

Backwater valve prevents
sewage contamination
of greywater.

Greywater diverter: 2" ABS
three-port valve

Cleanout before each
flow splitter

Double-ell flow splitter

Pipe end, protected by
valve box

Mulch basin

Figure 6. Diagram depicting a branched drain greywater reuse system. Drawing by
Leigh Jerrard.

political engagements emerging through Angelenos' entanglements with
such intimately scaled wastewater reuse infrastructures. Here, greywater
systems operate as sites where environmentalists can make claims to city
resources and advance new notions of both state and individual obligations to the management of the city's flows. Visibility plays a central role in
this process of engagement, as advocates have publicized certain forms of
illegal waterscape tinkering to contest the legitimacy of the state's claims
to regulate residential consumption and resource management in the
public's best interest. Attending to such practices, I use greywater systems
to explore how these nodes mediate some residents' demands for a more
extensive devolution of waterscape control within the city.

As Hank's remarks suggest, novel projects and understandings of care
for the urban landscape can arise (and be realized) through greywater
reuse. Redirecting their home wastewater from the grid to their gardens

Figure 7. Under-construction branched drain greywater reuse system. Photo by Leigh Jerrard.

Figure 8. Completed branched drain greywater reuse system. Photo by Leigh Jerrard.

spurs some people to both change their consumption habits and reassess the role of residents and domestic space within the metropolitan environment more generally. Complementary ethnographic research in the US Southwest documents similar shifts, illustrating how reusing greywater and capturing rainwater can reshape residents' understanding of themselves and the landscapes they inhabit.[3] Such accounts emphasize how, in subverting the modes of home-scale consumption quietly "baked in" to the water grid, these interactions with the material stuff of the home waterscape can serve to inculcate more attentive relations with both local and distant watersheds.[4] While decidedly quotidian, these shifts in practice and perspective suggest the fluidity of established urban norms of water consumption and environmental attunement.

These modes of attention also underpin greywater advocates' efforts to mobilize domestic-scale reuse systems as a particular kind of techno-political terrain: strategically visible nodes of engagement with the state, in the form of public water agencies and regulators. Beyond fostering new environmental practices and feelings within homes and yards, these infrastructures are also drawn into efforts to incite more engaged and critical stances toward the institutions tasked with managing the city's sprawling water network. Public agency workers' resistance to this form of selective disconnection from the grid clarifies the patterns of flows, norms of consumption, and forms of state control at stake in efforts to decentralize established arrangements of urban wastewater management. Tracking these conflicts also reveals the divergent notions of sustainable domestic water practice associated with the project of replumbing the city.

Considered in the context of the stormwater recharge projects explored over the three chapters that follow, the tensions around greywater systems elaborated here reveal an inconsistency. After all, the city actively promotes home-scale runoff infiltration infrastructures, particularly within the aquifer-connected northeastern Valley. Such distributed installations (and the work of maintaining them) are framed as key tools for transforming the city's rainwater from a hazard into a resource, buffering LA's water supply from the anticipated impacts of climate change. But when it comes to the city's sewage flows, public agencies resist such decentralized arrangements and active forms of resident engagement with intimately scaled water infrastructures. The divergent spatial logics that guide city approaches to these two water resources in the making illustrate contradictions that quietly structure state-led efforts to replumb LA. Lingering here on the forms of water management decisions and adaptation work that public agencies actively avoid devolving to residents highlights the possibility of far more extensively redistributed arrangements of waterscape labor within the city—and what, exactly, these institutions stand to cede by pursuing them.

DISCONNECTING FROM THE SEWER—AND FROM HYPERION

Not long after I first met Hank in the summer of 2012, he invited me to join a tour he was organizing of Hyperion, an enormous LA wastewater treatment facility separated from the Santa Monica Bay only by a road and a sandy strip of beach. At the site, an employee of the city's Bureau of Sanitation drove us between buildings in a clackety open-air cart, explaining technical processes and sharing trivia at each stop. (A representative morsel: scenes in the 1973 sci-fi classic *Soylent Green* were filmed at the plant.) Most of us, including Hank, took dozens of photos and videos of the plant's gigantic trash-removing screens, bulbous anaerobic digesters, and dark settling ponds. Over the years that followed, I noticed that several of those images found their way into Hank's standard introduction to greywater PowerPoint presentation. Displayed alongside photos of the aqueducts piping drinking water into LA, he used the snapshots to illustrate the hulking, relatively distant infrastructures to which residents' homes are connected via the city's water grid.

In both public presentations and private conversations, Hank was critical of the socio-ecological damage that this network has brought to a range of landscapes, water bodies, and communities, both within and beyond LA. His framing of Hyperion, as a place where the city relies on tremendous inputs of energy and chemicals to sort-of clean effluent that it then dumps into the ocean, was particularly harsh. Recounting such critical perspectives on the plant, Hank's comments reflected the writings of local historians who have chronicled the plant's contributions to poor water quality along LA's coastline.[5] Like many of my greywater advocate interlocutors, he invoked the plant and the LA Aqueduct—that is, the city's biggest, most visually prominent water infrastructures—to tell a reproachful story about its urban metabolism. Drawing attention to these relatively well-known sites within the extended urban waterscape to question and contest public agencies' stewardship of the city's flows, he expressed a sense of dissatisfaction unexceptional among environmentalists across the city.

Over time, however, I came to appreciate how the greywater community also used the fixtures associated with day-to-day water consumption to advance such critiques of the city's waterscape and its management. Toggling between massive facilities like Hyperion and ordinary bathroom taps and drains in their discussions of the urban water system, advocates sought to raise questions about the wisdom of allowing institutions with expertise in managing the former to dictate residents' day-to-day engagements with the latter.

When Issa and her husband decided to have a greywater system installed at their San Fernando Valley home, they were familiar with these critiques. They had, after all, sat through Hank's well-rehearsed greywater overview lecture at one of his company's Saturday morning installation workshops. A biracial woman in her thirties, born and raised in Northern California, Issa understood that most of the water streaming from her household tap in LA came from hundreds of miles away. For her, the appeal of home-scale reuse was largely about the opportunity to make better use of those flows than city agencies currently did. The waste of dumping all that water into the sea really got under her skin. That frustration had only sharpened since she had relocated to Southern California a few years earlier, particularly given all the worrying headlines about the drought.

Not long after that move, Issa and her husband bought a home and began to contemplate what to do with its yard. They knew that they wanted to remove the thirsty lawn—as she put it, "We have a responsibility not to live as though our yards are in England!" But now that they had a bit of space, they also wanted to produce food and maybe some pollinator habitat, using their slice of the urban fabric to improve LA's broader ecology. At first, anxiety about the associated water consumption led them to keep several buckets in their shower. The work of dutifully hauling the water-laden pails out to their fledgling garden and fruit trees led them to investigate what it would take to redirect some of their wastewater away from the sewer in a less labor-intensive manner. Hiring someone to install a greywater system, they eventually decided, wouldn't be cheap, but it would be the best long-term way to turn their laundry and shower effluent into liquid that served some purpose for their local environment. Maintaining the system would take some effort, but less than the constant bucket lugging, they ultimately concluded.

Issa had shared some of this story with me as we worked together digging trenches for greywater pipes during the workshop she attended. Months later, we met again at her house to discuss the reuse system that Hank's company had recently constructed on the property. Technically, she explained, there were two systems, one directing laundry effluent to the front yard and one directing shower water to the back yard. (This much I already knew from assisting with the installation. The experience had also taught me that her house's crawl space was too small to allow comfortable squatting, a fact that had complicated the job.) Both systems irrigated fruit trees. Their herb garden, planted in a raised bed, was still being watered from the bucket in which her husband rinsed vegetables before cooking. But overall, she told me, they felt much less worried and guilty about their modest day-to-day water consumption: "It's not just that we're not wasting the water. We wanted to *use* it." Laughing a little, she told me that before the recycling system was built, her husband had cut his showers down to twice a week—but now that the used water was routed to the garden, he was back to a near-daily schedule (and smelling a bit better).

Describing her home, yard, pipes, and daily practices in this manner, Issa was sketching a moral ecology—that is, her notion of "just relations between people, land, water, and nonhuman plants, buildings, technologies, and infrastructures."[6] The senses of anxiety, guilt, and responsibility she articulated in connection to the waters flowing through her home and garden signal the forms of morality she attached to this small slice of the LA landscape. Grounded in the conviction that wasting the water piped in from afar was an act of profligate entitlement, Issa understood her yard of effluent-sustained herbs and fruit trees as a configuration that furthered collective flourishing both within and beyond her property lines. In such accounts, a "good" yard emerges as a classification rooted in its material inputs and perceived ecological effects rather than just its aesthetics, a growing perspective among middle-class Americans.[7]

Resonant accounts of the moral and ecological stakes of home-scale consumption, waste, and land cultivation marked many of my conversations with greywater adoptees. I first encountered Emma, a white woman and a retired small business owner, when she took the microphone during the comment period at a public meeting about a stormwater planning document. Held on a weekday morning in an auditorium at the LA zoo, the forum was dominated by water consultants and staffers from city agencies and NGOs. When addressing the gathering, Emma spoke forcefully, roasting the plan's authors for ignoring the robust role that individual homeowners could play in transforming LA's water system. "I put in swales and cisterns and even a greywater system on my own home. It wasn't that hard and other people should be doing this, too!" she told the crowd. "I've been saying this in public meetings for years now, and the planners need to listen!" When she finished, the forum leader politely thanked her for the comments and her personal water savings, seemingly taken aback by her forceful dismissal of the plans he had termed "aggressive" and "groundbreaking."

A few weeks after that event, I visited Emma's westside home for a conversation about greywater. No less direct or confident in that setting, she proudly marched me through a pair of tours: one of the elaborate water capture and recycling infrastructure she'd installed around her home, and another of her years of public advocacy for more sustainable water policy and infrastructure. For her, the combination of making material changes to her home waterscape and pushing for policy change was a necessary pairing. "Anyone who works with nonprofits knows that democracy is a slow, lumbering animal, and I'm not," she explained. "I need to do something in addition to the democratic process." Installing a greywater system allowed her to feel like she was making a small but beneficial difference within the urban environment in the immediate term, even if her efforts to push the city bureaucracy toward a more sustainable water provision system were often stymied. Living with the reuse technology helped sustain her desire to transform the city's water grid on a larger scale.

Omar, a onetime client of Hank's, took a different line. A homeowner in his mid-thirties living in an upscale section of the San Fernando Valley,

Omar was a dentist and a self-proclaimed "regular guy"—albeit one who had retrofitted his house with the most advanced water capture and reuse systems I saw in my travels across the city. When I arrived for our interview, he also took me on a lengthy walkthrough of his carefully decorated home, stopping to admire the rain tanks and greywater system and solar panels. When I mentioned that he seemed well-rehearsed in covering this circuit, he explained why: he gave frequent tours of the place to various local environmental groups and neighbors and friends. His goal, he told me, was to prove that you can have a super-light ecological footprint without being a hippie, without being fringe. This was the reason he wouldn't install a composting toilet, an act he saw as the final step in water savings. "I've got to date," he explained. "Women will think I'm weird if I have something like that."

As part of his project to better "mainstream" ecological consciousness, he had taken to making and posting videos on YouTube. He described his favorite, which drew on familiar tropes of "leaving the closet" narratives to tell the story of a West Asian immigrant like himself "coming out" to his parents—but coming out as an environmentalist, rather than revealing a concealed sexual orientation or identity. "My parents think this is all nonsense," Omar explained. "Frivolous, a waste of time. . . . [T]hey wouldn't take out the lawn at their house until I got them the lawn rebate and showed them it made financial sense." He admitted that his retrofits were not cheap, and that this scale of remaking the home is not available to everyone. But he knew people who had the money, and he thought that by making himself an example, he could nudge them into making changes to cut their own consumption.

As these stories attest, for many greywater system adoptees, living with infrastructure that redirects domestic effluent out of the water grid and into the urban landscape can be an arrangement thick with meaning. Much like other sustained projects involving embodied forms of environmental tending, the everyday practices that such fixtures require to remain functional, along with the social exertions of promoting them to friends, family members, YouTube viewers, and auditoriums full of engineers, can contribute to new forms of environmental subjectivity.[8] But while taking such orientations and relations seriously, we do well to situate them in the socioeconomic context within which they are often rooted.

Hank's clients ranged from comfortably middle class to very wealthy. While some discussed their choice to install a greywater system in terms of potential savings on their water bills, most acknowledged that it would likely take decades for such reductions to exceed the cost of system installation. For the majority, these systems were understood as optional home upgrades that moderated their ambient guilt associated with a relatively resource-intensive lifestyle. As such, these affectively loaded accounts of rerouting domestic flows should also be understood as, in many cases, articulations of a local strain of bourgeois environmentalism inaccessible to many within the city.[9]

Even so, as the section that follows details, in the context of Southern California, greywater reuse has been and remains a somewhat controversial act, due to the ways in which it disrupts the spatial and governmental logics of urban water networks. Home-scale greywater systems have been a source of friction and the site of active negotiation between greywater advocates and water agencies, nodes where practitioners have challenged subtle forms of state control within the water grid.

REGULATORY EVOLUTION AND STRATEGIC GREYWATER VISIBILITY

Founded on the edge of LA's Koreatown neighborhood in 1993, the LA Eco-Village has long been a hub for local activists of many stripes. An intentional community with a few dozen members, the Eco-Village frequently hosts environmentalist talks, movie screenings, and workshops open to nonmembers. Most residents inhabit a small cluster of aging apartment buildings arranged around an internal courtyard, where groups of mismatched chairs share space with a patchwork of garden plots. On a gray June morning I joined about a dozen visitors in the Eco-Village's second floor community room for the first day of a weeklong greywater installers' training course. Following a brief get-to-know-you exercise, which required us to find fellow course enrollees who identified with phrases from a list of descriptions (including "loves to grow her own food," "practices permaculture," and "uses less than twenty-five gallons of water a day"), we settled in for an introductory lecture from Ana, our primary instructor. A white woman in her thirties, Ana presented a pithy talk grounded in years of both system installation and greywater advocacy, a combination of experiences she threaded together seamlessly in her teaching.

Near the beginning of the lecture, she offered an overview of how installers understand the flows from different domestic fixtures. Laundry wash water is a great source for greywater, she told us, as long as you use certain kinds of detergent, and ditto for shower or bathwater if you're mindful about your soaps and avoid bath salts. Toilet flows she called "blackwater," describing them as strictly off-limits for on-site reuse due to safety issues. Kitchen sink effluent she characterized as something in between: "I call it dark greywater. There are safe ways to reuse it, but you have to be very careful. It's pre-legal, but we're working on that with the regulators."

Discussing greywater regulation as a malleable set of rules, Ana spoke from experience. Through more than a decade of advocacy, she had helped shift the status of on-site reuse from almost entirely illegal to largely allowable under the California Plumbing Code, and even more acceptable in several city building codes within the state, including LA's. Tracing this arc helps to clarify how greywater systems have served as nodes where advocates have steadily challenged both norms of domestic water use and public agencies' legitimacy to govern the home waterscape. Pursuing strategic forms of greywater system visibility has played a central role in these fixtures' use as such a techno-political terrain.

A 2007 *New York Times* article titled "The Dirty Water Underground" serves as an instructive example of such tactics. The reporter describes the home of Laura Allen, a member of an Oakland-based group called the Greywater Guerrillas, in the following terms: "Jerry-built pipes protrude at odd angles from the back and sides of the nearly century-old house, running into a cascading series of bathtubs lined with gravel and cattails. White PVC pipe, buckets, milk crates, and hoses are strewn about the lot. Inside there is mysterious—and illegal—plumbing in every room."[10] Photos of the self-described Guerrillas and their code-noncompliant home plumbing are interspersed within the story, which details how these advocates endeavor to spread such systems among California residents, demonstrate the efficacy of domestic reuse as a water conservation technique, and transform the regulatory codes that forbid such practices. Challenging both the authority and the logic of the laws governing home-scale reuse, the advocates contest the unmarked forms of state power that structure day-to-day interactions with the domestic waterscape.[11] As the choice to participate in the *Times* article suggests, the work of making home-scale greywater systems visible and legible to both regulators and a broader public has been a defining element of these engagements.

For decades, these efforts were directed toward combating state assertions that home-scale reuse poses a threat to human and environmental health. A 1980 California Department of Water Resources (DWR) report titled *Residential Greywater Management* describes greywater use that originated during the state's 1976–1977 drought as a novel practice, one necessitating a comprehensive review to aid the development of appropriate regulatory guidance.[12] At that point, greywater recycling was regulated only at the county level and outlawed in all fifty-eight California counties on the grounds that reusing waste effluent in the yard had potential public health risks. Yet despite a smattering of additional reports from the DWR, most of which took a cautiously optimistic tone toward the safety of reusing greywater on-site while delineating knowledge gaps related to pathogen transmission, both research on the actual health risks associated with greywater and a state-scale regulatory framework were slow to emerge. When California's next big drought took hold in the late 1980s, greywater's legal status remained largely unchanged: prohibited in all counties on safety grounds, with no state-level code or guidance framing local policy.

Greywater advocates in LA and beyond, cognizant of the role that such official forms of ignorance played in the technique's continued marginalization, pushed public agencies to produce knowledge about the technique and its safety. In 1990 an environmentalist city councilwoman directed LA's temporary Office of Water Reclamation to conduct a small pilot study to gather credible information on the risks of exposure that the systems present to users. Though still technically illegal, eight reuse systems were installed in LA homes for monitoring. The study's results suggested that these residents need not worry overmuch about contamination associated with these systems, because it showed that a lot of LA dirt contains concerning contaminants, whether or not one irrigates it with greywater.[13]

Following that report's primary policy recommendation, in 1994 the city approved an ordinance to permit greywater systems within LA, mirroring the trajectory of state-scale greywater regulation. In 1993, in response to drought-time environmentalist pressure, the California State Assembly unanimously approved AB 3518, a bill instructing the DWR to develop statewide guidelines for irrigation with residential greywater. On January 1, 1994, an updated California Plumbing Code section permitting limited greywater reuse for on-site landscape irrigation became law. While cities and counties retained the capability to enact more stringent regulations to manage the resource, one major roadblock to wider implementation of distributed reuse seemed to be gone.

Yet when people tried to build permittable, code-compliant greywater systems, they quickly discovered that to minimize risk, under the new rules all greywater was to be treated like septic water. The technical specifications of the code required deep underground disposal, involving a tank and gravel-filled leach lines. In response to this onerous, expensive, hyperprescriptive approach to laundry water and bathtub effluent, greywater advocates began to construct and promote simpler, illegal systems, explicitly challenging assumptions about the dangers of reencountering shower or laundry water that structured the guidelines.

The Greywater Guerrillas were at the forefront of this work during the 2000s, and their influence spread across the state. In addition to installing systems and holding daylong workshops, their members hosted weeklong greywater installer training courses that drew Angelenos, including several of my interlocutors, looking to learn more about both greywater plumbing and its regulation. Members also published *The Greywater Guerrilla Girls Guide to Water*, a promotional zine that they passed out for free, and *Dam Nation: Dispatches from the Water Underground*, an edited collection of essays, photos, and cartoon diagrams outlining the group's critical assessment of water management arrangements across the region.[14] In these works, the Guerrillas emphasized the code-noncompliant nature of their practices alongside the simplicity, safety, and efficacy of their home-scale reuse systems in reducing water consumption.

Flouting the codes and flaunting their illegal systems while foregrounding the safety and water-saving potential of on-site reuse, the Guerrillas directly challenged public agencies' competence to govern home-scale water reuse under conditions of growing water stress. This sustained pressure paid off during California's 2007–2009 drought, when public agencies faced heightened scrutiny for constraining domestic conservation. Working directly with state regulators, the Guerrillas played a central role in overhauling the California Plumbing Code. Under the updated rules, which came into effect in 2010, mandatory leach lines were a thing of the past. If greywater outlets were buried at least two inches below the landscape's surface, system designers had relatively free rein on their form. Simple laundry-to-landscape systems now required no permits to be legally installed.[15] After the new code passed, the Guerrillas changed their organization's name to Greywater Action, to acknowledge the state's move to effectively legalize their long-standing practices.

As Ana's description of the "dark gray" kitchen sink water during our installers training suggests, advocates have continued to push regulators to classify more forms of domestic water as safe for on-site reuse, bringing regulations closer to alignment with on-the-ground practice. And formal knowledge production regarding the impacts of systems on the local environment remains central to this work. In 2013, Greywater Action released a systematic study of the impacts of eighty-three residential greywater systems on soil quality, water consumption patterns, user satisfaction, and installation costs, work that, as I observed, LA advocates cited frequently as evidence of the safety of domestic reuse practices.[16] But as the next sections detail, disagreements over questions of sustainability and socio-ecological benefits, rather than those of safety, now shape the most contentious negotiations between greywater advocates and public agencies. As I demonstrate in the following pages, these frictions are rooted in conflicting understandings of the appropriate role for the city's residents within the waterscape.

PUBLIC AGENCY RESISTANCE AND THE ROLE OF THE WATER USER

Near the end of a chatty lunch in a downtown LA food court, an engineering consultant named Iris asked me to explain something she couldn't quite grasp: "Why, exactly, do people like greywater?" Taking in my confused expression, she explained that the idea of many Angelenos installing home greywater systems made her worry. She noted that, thanks to the spread of more-efficient indoor water fixtures and conservation initiatives, the city's sewer flows are already down from volumes typical of the 1980s. "If everyone diverts their shower and laundry water to their gardens, then we'll just have less for the centralized wastewater treatment plants to clean," she reasoned. "Which wouldn't be a great situation if LA were to invest several hundred million dollars into the infrastructure to clean those flows to drinking water standards and recycle them back into the city's potable supply, right?" When I suggested that some people find it empowering to reuse their wastewater on-site and consider it a meaningful way to nurture the local environment, she frowned and began talking about the challenges of proper public water education: "We need to figure out a good way to teach people that letting water go to the sewer can be a better form of taking action and helping solve the water problem."

Presenting greywater reuse in such dismissive terms, Iris echoed the opinions expressed by other public agency interlocutors over the course of my fieldwork. In these accounts, widespread residential greywater reuse threatens to undermine more efficient and dependable municipal-scale water recycling projects. Put another way, the home-scale systems serve as nodes where ill-informed users can complicate the "real" work of climate proofing the city's water system. Though quick to acknowledge that large-scale recycling infrastructures have for decades remained plans rather than functioning plants, these managers are steadfast in their contention that centralized wastewater recycling is a far more "sustainable" direction for LA's waterscape than domestic reuse. Beyond noting that definitions of urban sustainability are frequently contested, in this context it is helpful to consider what exactly is being sustained in arrangements that bear this label.[17] Here, I observed that a particular understanding of the residential water user, grounded in the state's established role in overseeing the urban water grid, is carefully preserved within this vision of the sustainable, replumbed city.

Slippage between material and social concerns often marked my water agency interlocutors' critical accounts of home-scale reuse. Questions about the efficacy of domestic water recycling as a water management practice were frequently elided with those about the capacity of residents to sustain such efforts for a protracted period. As with Iris, resistance to greywater system uptake was typically expressed as both a volumetric and a planning issue: too much domestic wastewater reuse would decrease flows through the sanitary sewer system, complicating the work of designing an appropriately scaled municipal reuse plant and potentially undermining its contribution to the city's potable water supply. In the words of Anne, a LADWP staffer, "From the city's perspective, we're making a huge investment in advanced treatment . . . and now everyone's going to do a greywater system? That makes their treatment job a lot harder. The systems are designed for a certain amount of flow, and because of that, I see some heels digging in."

Anne's next comments further developed this sense of anxiety about the uncertainties of greywater, moving from a focus on wastewater volumes to an assessment of LA water users: "I'm torn about decentralized solutions like greywater. Someone can be a total greywater junkie, but then they sell their house—and houses here are only owned for an average of eight years—then someone inherits the system, and the person taking it on doesn't give a shit. . . . There are efficiencies to centralized systems, that's why they're built." Here, greywater systems are presented as remarkably fluid infrastructures, their efficacy subject to the vagaries of housing stock turnover and the whims of future homeowners who do not care about maintaining the material stuff of their inherited systems. The implicit contrast drawn—between the dependable management of centralized infrastructures, carried out by the steady government, and the uncertain oversight of decentralized systems, undertaken by capricious individual water users—suggests that trusting LA residents with the serious work of managing water or water infrastructure is not desirable. Similar perspectives surfaced in almost all my interviews with water agency workers, many of whom described greywater reuse as a trend that emerges during droughts and disappears when the rains return. Water users, in these accounts, were simply not reliable.

Such a pessimistic understanding of Angelenos aligns with the city's dominant conservation messaging and programming during my fieldwork period. Starting in the middle of 2015, LA paid to plaster banners on billboards, benches, and city buses exhorting residents to more carefully monitor their consumption practices. This "Save the Drop" campaign featured a big-eyed blue water droplet offering factoids on topics like how much water residents waste when they let the faucet run while brushing their teeth, imploring them to engage more mindfully with their taps.

The figure of the unthinking LA water user also took center stage at a Water Hackathon cosponsored by the LADWP and a local tech company in summer 2015. During a break between presentations, I struck up a conversation with Yousef, one of the participants. His team of mechanical engineers had developed a prototype sensor that one could place by a household faucet. Using the sound waves produced by water flowing from the tap, the device measured the volume of liquid pouring out during use. Set to a target volume for daily water use, the sensor would start the day emitting a bright light, which faded gradually as the household approached its target. The goal of the device, Yousef explained, was to make people aware of their rate of water consumption as they're consuming. "Because right now, even though we know that we should conserve, we don't know how much water is flowing!" he exclaimed. "This lack of information and feedback is the biggest impediment to behavior change today." Most of the other "hacks" I learned about during the event operated on a similar logic: water users need to be made more aware of their water usage to enable them to reduce it. Drought-shaming maps that identified outdoor over-waterers, other faucet flow measuring devices, and simple "smart" meters were among the solutions proposed to solve the problem of mindless, profligate water consumption behaviors. In effect, the water agency's version of an LA water user inhabits the abstracted world of expert-generated sustainability metrics and spaces of calculability and can do the most good within the water system by moderating a prescribed form of consumption behavior.[18]

Scholars writing in other contexts that approximate the modern infrastructural ideal of universal, publicly provided water and sewerage services have highlighted the work that discourses narrowly focused on the figure of the unthinking water user can do to obscure the roles of public infrastructural networks, cultural and aesthetic norms, and water agency billing arrangements in producing such wasteful practices.[19] Here, I emphasize this figure to foreground the efficacy it quietly attributes to institutions managing the larger-scale elements of the water grid. Iris's contention that letting water run to the sewer helps to solve the city's water problems is premised on the notion that public agencies will steward the resource responsibly and effectively. Following Ormerod's framing of such perspectives as neosanitarian (discussed at greater length in the previous chapter), I contend that such an approach entails continuity with the robust modes of state governance and control within the waterscape that are associated with the modern infrastructural ideal.[20] To maintain the unreliability of LA water users and resist domestic greywater reuse is to help preserve the state's role as the legitimate manager of the city's water grid, competent in stewarding the network into a climate-changed future. To accept residents as potentially capable managers of and decision-makers regarding their wastewater would not only threaten the flow of effluent toward centralized facilities; it would also serve to undermine such technocratic authority.[21]

The next section elaborates the mismatch between water managers' account of the mindless LA water user and the greywater advocates' understanding of the same figure. In the advocates' formulations, greywater systems serve not just as sites where residents take on some of the work of urban water management and learn to understand the boundaries of the water network differently but also as nodes where oppositional engagements with the agencies overseeing that infrastructural system can be cultivated.

GREYWATER PRACTICES AS POLITICAL CONDUITS

One Sunday afternoon in early November 2014 I drove a car full of greywater advocates north from LA to a conference, hosted in a resort facility just outside of Yosemite National Park. During our dinner break in a roadside Mexican restaurant, Tom, a white man in his sixties and a longtime greywater advocate, spent some time explaining to me why he sees greywater as such valuable technology: "It's not just about making people use less water—it's also about getting them to see flows. It's like a good gateway drug to get people thinking about their homes and yards holistically and systemically. It eliminates the notion of 'away'—you start to think about how much water your family consumes and the personal care products you use, and how those volumes and compounds will affect what you can grow in your yard, and where you can grow it." Speaking in these terms, Tom was describing greywater reuse as a practice that reworks not only material flows but also adoptees' environmental consciousness, inculcating new forms of attunement to and care for the local landscape.

In his remarks at the conference a few days later, Tom pushed these ideas even further. The overarching goal of his career promoting greywater, he told the assembled crowd, had been to use the systems to make people more like "terrestrial beavers," citizens of the earth who improve the overall health of the ecosystems they rely on rather than just sucking resources from them. Tom's beaver comment was locally resonant, as the night before a biologist had shown up at the conference's welcome party wearing a plush beaver costume and performed a song about the benefits beaver dams bring to watersheds. It was also a gesture toward a perspective that animates the work of many of my greywater advocate interlocutors. Reusing greywater, I was often told, is a meaningfully different practice from simply working to use less potable water in and around the home. Watering your landscape with greywater allows you to transform that substance from a waste to be disposed of into a source of benefits for your local environment. And with the right encouragement, cultivating a new resource in this manner can seed critical engagements with the entities who manage a grid and make such reuse unnecessarily difficult.

Assumed within such formulations is a sense of urban residents' obligation to nurture the landscapes they live within and an understanding of domestic greywater reuse as a form of social reproduction of the extended urban environment.[22] Tom and his colleagues maintain that through quotidian acts of attention—washing with landscape-friendly shampoos, regularly scooping out gunk from the greywater system's mulch basins, carefully maintaining the plants watered with effluent—residents establish caring relationships with a localized, more-than-human community that enable them to fulfill their responsibilities to said collective. Within such a perspective, greywater systems, by linking familiar home water fixtures to the urban landscape in new ways, effectively extend the water grid's possible range of ecological benefits and expand the potentially salutary role of the resident within urban water management. Robust notions of extended connection and obligation underpin these socio-ecological imaginaries.

Related research has questioned the transformative potential of domestic infrastructures and the practices they demand, troubling assumptions that a growing sense of care for the local environment necessarily entails a justice-oriented perspective. Anthropologist Michael Vine's work on Southern California desert residents who have reworked their home water infrastructures by installing greywater systems and composting toilets is useful in its emphasis on both the subject-making function of these "everyday experiments."[23] Reflecting on the settler-colonial mindsets he observed being reproduced through these efforts, Vine writes, "The everyday experiments examined here reinvigorate libertarian ideals of the rugged self-sufficiency on an imagined 'frontier,' working to replace more interconnected modes of ecopolitical activism with the figure of the atomistic individual located in private space."[24]

These findings emphasize the limitations of home-scale environmentalist practices as a basis for an environmental politics that grapples with racialized forms of environmental inequality and systems that (re)produce such conditions—in contrast to, say, engagements grounded in efforts to address the uneven distribution of environmental harms and benefits within the city.[25] Such points are highly salient here, particularly among greywater advocates, a group that skews whiter and wealthier than their city as a whole (albeit less so than the system adoptees that I met), and that often targets promotional attention toward homeowners, given the limited latitude that renters have to cut into home plumbing. However, within this context, the advocates also seek to unsettle the very notion of the "atomistic individual located in private space" by foregrounding the material connections between the domestic sphere, the public water grid, and the landscapes produced via the city's metabolism.

Such links—between the home, the city's large-scale water infrastructures, and the urban and distant landscapes connected by the water network—featured prominently in Hank's initial consultations with homeowners. For these meetings, we visited the houses of prospective clients to figure out what kind of greywater system would be feasible to install on their property. Roughly one-half site assessment and one-half sales pitch, these discussions usually began with Hank, me, and the homeowner seated around a kitchen table, poring over Hank's binder full of photos. Pausing on images of his front yard lemon trees, Hank would detail the local environmental benefits of installing a greywater system, often referencing the spate of drought-related tree deaths that LA was experiencing at that time. He would move on to explain how applying used wash water on the landscape not only helps the plants you intend to water but also contributes to the overall level of the water table without using potable water—a benefit, he observed, that was not realized when that effluent was routed into the sewers.[26] Notably, Hank's binder also included images of the pumps that lift water over mountains to get it to Southern California, as well some snapshots from his Hyperion tour. While he talked a lot about very local environmental impacts, he also took pains to place the systems within the large, energy-intensive infrastructural network that moves and cleans water within California, as well as its negative impacts on both Owens Valley and the Santa Monica Bay.

An emphasis on connections between the domestic scale and the broader waterscape marked many of my conversations with greywater advocates, who tend to attribute a powerful political valence to the modes of understanding and participating in the waterscape that adopting this technology inculcates. Scholar Cleo Woelfle-Hazard, cofounder of the Greywater Guerrillas, offers a particularly clear articulation of this perspective in the introduction to the group's *Dam Nation* anthology: "People like greywater treatment wetlands—bathtubs full of gravel and cattails—because they're an example of a human-scale answer to the problems of centralized sewage treatment and mega-dams. If everyone built one in their backyard, would that bring the dams down? No. But guerrilla plumbers profiled in this book say that the taste of autonomy that building these systems provides has gotten them to question the logic of the water grid" (Woelfle-Erskine et al. 2006, 6). Woelfle-Hazard goes on to suggest that questioning the grid's logic is often connected to more active, combative participation in waterscape governance, in the form of protesting new water transference projects, critiquing the impacts of existing pipelines, demanding dam teardowns, and advocating for water providers to support the expansions of other forms of distributed water provision, such as rain gardens and cisterns. Understanding that they can change a small part of the waterscape (by rerouting their wastewater flows) makes the whole system appear more mutable—and worth actively contesting.

This framing echoes the writings of Kaika, who suggests the politically transformative potential of water users rethinking the domestic waterscape in contexts that approximate the modern infrastructural ideal. Like

Woelfle-Hazard, Kaika contends that such engagements can engender a different kind of political consciousness, one more attuned to the material connections between relatively far-flung spaces: "Demonstrating the ideological construction of private spaces as autonomous and disconnected and insisting on their material and social connections calls for an end to individualization, fragmentation, and disconnectedness that are looked for within the bliss of one's home. It calls for engaging in political and social action which is, almost invariably, decidedly public" (Kaika 2005, 75). Denaturalizing networks of provision is, in both cases, framed as the key first step toward imagining and demanding another set of material arrangements. Given this analysis, we might characterize greywater boosters as understanding greywater systems as a tool to promote an ongoing exploration of the "uncanny materiality" of pipes and flows within private homes. Beyond helping everyday urban residents become restorative "beaver"-type presences within their immediate environments, they also hope to inculcate a grounded understanding of—and a new sense of obligation to—the broader, shared waterscape that they inhabit.

To be clear, none of the advocates understood this trajectory as inevitable. Emma, the firebrand retiree demanding more aggressive stormwater capture investments from city agencies at the public forum, was not assumed to be a representative greywater adoptee. Many people, it was acknowledged, would continue to scope their ecological efforts at the scale of their property after installing a system, at most hoping to inspire a few friends to eventually follow their lead. For instance, Issa, so committed to reuse and her herb garden that she augmented her greywater system's outflows with hand-hauled buckets of vegetable wash water, answered with a flat no when I asked her if she was interested in any form of environmental activism or direct critical engagement with city agencies. She hoped that her household's approach to urban water and land use would eventually become mainstream, but she had no interest in pressuring anyone to make that happen.

Some adoptees, however, proved willing to contribute to efforts to lean on the LADWP and other public agencies to reconsider policies and funding priorities, challenging the ecologically damaging infrastructural logics and effects that Hank would criticize in his introductory presentations.

Figure 9. Greywater workshop participants laying effluent piping in a garden. Photo by author.

During my fieldwork, I observed advocates working to mobilize greywater adoptees to engage directly with their water agencies and to demand such changes. From late 2014 through 2015, local advocates curated, hosted, and promoted a series of Greywater Alliance meetings, held in a LADWP building near downtown LA. These sessions were developed to establish an active dialogue between public agency workers, regulators, greywater practitioners, greywater adoptees, and other water-concerned residents, and to push the agencies to take a more aggressive approach to promoting urban water conservation during the drought. In particular, these advocates sought to extract a rebate for residents who submitted proof of installing a home greywater system, to match the financial incentives the city had adopted for residents who purchased rain tanks or ripped out their turfgrass lawns. As Ana explained to me after one such meeting, the gatherings were modeled on a similar initiative in the San Francisco Bay Area, which had led to both shifts in greywater regulation and water agencies offering financial incentives for customers to reuse their wastewater on-site.

In LA, these efforts were somewhat less successful. While LADWP staff attended the Alliance meetings, they typically sat at the back of the room and spent much of the time attending to their phones. Within the city, the rebate program advocates hoped for did not materialize. As such outcomes suggest, increased (and increasingly combative) public engagement with water management agencies is not necessarily correlated with quick shifts in public policy. Nevertheless, such efforts underline the fact that greywater systems here are used to incite not only new consumption practices and localized affective attachments but also new forms of oppositional engagement with public water agencies, rooted in an expansive understanding of residents' capacity to contribute effectively to the management of the urban waterscape. Enrolled in advocates' efforts to enact a broader transformation of LA's water grid, these fixtures serve as sites where established logics and practices of state-led water management are actively contested.

As detailed across the past two chapters, the need for LA to make better use of urban wastewater flows is a point on which environmental advocates and water agency workers largely agree. My interlocutors from both groups emphasized the looming threat that climate change poses to the water supply of this sprawling city, situated as it is within a water provision system facing increasingly variable precipitation patterns. All understood the city's existing water provision arrangements to be increasingly precarious and described the city's sewer flows as an underutilized waste stream that should serve more functions within the urban landscape. The fight over greywater systems thus also helps to illustrate the frictions inherent in enacting forms of urban resource circularity.[27] Resisting the localized loops of home-scale reuse in favor of centralized, eco-modernist recycling facilities, public agencies perform an act of "enclosure" that quietly sustains established modes of consumption and techno-politics.[28] Doing so, they seek to realize an eco-modernist vision of climate adaptation that assumes continuity with the modern infrastructural ideal and the forms of state control within the waterscape that it entails. As such, greywater systems demonstrate that while state-led efforts to replumb the city's supply can draw urban landscapes and residents into the process of water provision in new ways (as detailed in the chapters that follow), they generally do not seek to displace established water governance arrangements. By and large, these projects seek to reroute LA's metabolism without troubling the institutions and relations that underpin it.

Approaching such infrastructures as techno-political terrain, greywater advocates have long challenged this paradigm. Through strategic mobilizations of home-scale reuse systems—as both sites of value-laden environmental production and nodes for engagement with public agencies—advocates aim to unsettle widely shared assumptions about the appropriate roles for both the state and individual residents within the urban waterscape. Contesting the best scale of urban water recycling, greywater advocates and water agency workers are locked in conflict over both the path of the city's effluent flows and the distribution of power within the waterscape.

Tracking how greywater systems have served as sites where advocates denaturalize established consumption norms and forms of state control within LA's water network helps to clarify how domestic fixtures come to be enrolled in a broader politics of urban metabolism, in a context where the provision network approximates the modern infrastructural ideal. As in settings where questions of political belonging are mediated via techno-political engagements with grids that provide vital resources to limited sections of the metropolitan fabric, here advocates tinker and heckle in the name of transforming not only individual practices but also the configurations of flows and control within the metropolitan waterscape. Such adaptation work is premised on the notion that a more distributed arrangement of water management, control, and work could serve as the basis for a less damaging network of flows.

The contrast between these skirmishes and the stormwater recharge projects explored in the chapters that follow is striking. As with greywater reuse, those projects rely on retrofitting and long-term maintenance work within residential landscapes to reroute water. But they diverge on a key register. Greywater systems, scattered throughout the city, remove already-contained liquid from the public water grid, whereas recharge projects, concentrated above target aquifers, seek to direct dispersed flows toward a common subterranean water stockpile. In effect, the city aims to cultivate the work of landscapes and residents who might help direct new flows of stormwater into the centralized network, but not those who would remove waste flows from that system (even if in the name of conservation and local ecological benefits). This distinction suggests the diffuse forms of control that the state seeks to preserve while adapting the urban waterscape to a climate-changed hydrology, along with the spatially and socially differentiated nature of this approach to urban resource security.

PART II　Recharge

3 Managing Landscape

RECHARGE IN THE NORTHEASTERN
SAN FERNANDO VALLEY

"It wasn't until I joined the department that I realized the San Fernando Valley has an aquifer," Luis told me with a rueful smile. As our months meeting at various city stormwater planning events had made clear, times had changed for him in this regard. During my fieldwork period, Luis was a midlevel LADWP engineer, spending a good chunk of his working hours helping to develop the agency's first-ever Stormwater Capture Master Plan (SCMP). While much of his effort focused on modeling the way that runoff moves across the city's surface, the aquifer sitting beneath the urban landscape was never far from his mind. Luis's modeling was, after all, carried out in the service of designing infrastructures that could help direct more stormwater into the San Fernando Groundwater Basin for storage and eventual incorporation into the city's potable water system.

As he explained in an interview at the LADWP's downtown headquarters, this orientation was based on a particular understanding of how the aquifer should function within LA's sprawling waterscape. Citing the ongoing drought, the strain that the dry spell was placing on the city's distant water sources, and the likelihood that climate change would make such conditions more common in the future, Luis suggested that a best-case scenario for LA would involve the San Fernando Groundwater Basin serving as a space where the city's stormwater accumulates, for use in case of shocks elsewhere in the system. He mentioned the aquifer's enormous unused storage capacity, currently estimated at around 550,000 acre-feet—more than the city's average annual consumption between 2016 and 2020—and reminded me that, unlike dam-contained reservoirs, filling it would not remove large swathes of land from development through permanent flooding.[1] Effectively, a fluke of geology had provided the city with an accessible water container larger than any surface reservoir in Southern California. As Luis put it: "For us, that's a storage tank that's sitting empty—that we need to send more water into."

Such conversations signal that, like LA's sewage flows, the stormwater that runs through city streets after a downpour is understood by city water managers as a water resource in the making and a key element of its approach to climate adaptation.[2] But in contrast to municipal wastewater recycling projects, which rely on liquid already directed to sewage treatment plants via a network of underground pipes, these engineers' work with rainwater entails a large-scale redirection of the city's runoff flows into the waiting "storage tank." To capture that water, they seek to make the land that overlies the San Fernando Groundwater Basin more permeable.

In public presentations made during the SCMP's development process, Luis and his colleagues used a pair of aerial photos of the northeastern San Fernando Valley to convey this orientation. The first, a black-and-white image taken in 1949, shows several winding watercourses and many large, undeveloped lots. In contrast, a colorful picture from 2008 captures a much denser pattern of development, the extensively subdivided arrangements of buildings and roads typical of urbanized Southern California. The streams in this image are straighter and narrower, channelized with concrete.

Explaining the two photos, the staffer presenting would tell the assembled crowd that the changes depicted in them have had major consequences for the aquifer lurking beneath this landscape. Fields and farms have become subdivisions and shopping centers and parking lots and freeways, which means that impervious surfaces like roof tiles and asphalt cover previously porous ground. Even the area's streambeds have been hardscaped. The problem with these developments, per the presenters, is that the Valley's groundwater basin is no longer getting replenished the

Figure 10. Stormwater glistening on a street in the northeastern Valley. The widespread introduction of impervious surfaces like asphalt has dramatically reduced the landscape's recharge capacity over the past century. Photo by author.

way it used to, because more stormwater is sliding into the city's storm drains and out to the ocean with every rainfall.[3] The purpose of the SCMP, they would then tell us, was to guide the process of engineering that lost absorbent capacity (and then some) back into these lands. And as the PowerPoint slides that followed would show, this transition was understood to entail developing stormwater recharge infrastructures in areas currently occupied by parks, power line easements, medians, parkway strips, and residential yards across aquifer-connected sections of the northeastern Valley. A carefully designed collection of swales and drywells and recharge basins would thus transform the landscape into infrastructural nature while largely preserving established land uses.

These presentations communicated a particular role for this section of LA within the urban waterscape: through dispersed infrastructural interventions, the area would be managed to soak in and hold runoff for the city. This characterization echoes Ellen's account in the opening pages of this book, as we crouched by the gravel pit and she described the northeastern Valley as an area of tremendous (but woefully underutilized) recharge potential. Such interpretations of this land are rooted in an assessment of a subset of its material characteristics: the rocky, porous soils that enable easy percolation of runoff; the massive volumes of stormwater that roll into it from the surrounding hills; and the partially filled groundwater basin lurking beneath its surface. In other words, these plans for the landscape's future aim to capitalize on its intrinsic hydrological and geological features, intensifying their water-corralling capacity. As such, the process of stormwater's resource becoming is premised on the transformation of the terrain of the northeastern Valley into a particular sort of infrastructural nature, land managed to maximize a selected set of ecological functions in the name of shoring up citywide water reliability.

The next three chapters explore various aspects of this replumbing project, tracking the relations and frictions entangled in its pursuit. Building on ethnographic considerations of infrastructural nature from other contexts, I attend carefully to the multiple forms of material and discursive work that realizing these arrangements entails.[4] These are efforts that rely on selective narratives of both other-than-human nature and people's capacity for its management—namely, that the natural world has the intrinsic capacity to provide human-selected functions, and that people are equipped to produce such rationalized ecologies. Such undertakings typically require the active reconfiguration of landscapes, as the ability to provide chosen ecosystem services must, as anthropologist Ashley Carse notes, "be built, invested in, made functional, and managed."[5] As the need for such forms of human labor and resources suggests, infrastructural natures can be understood as novel, hybrid ecologies, which intertwine people and capital with land in new ways. Proceeding from the view that the pursuit of such arrangements is best approached as an inherently political project of both reframing and reordering terrain, in the pages that follow I examine the halting process of envisioning and producing a distinctly urban form of infrastructural nature in the northeastern Valley.

The present chapter considers how the notion of managing this landscape as a space of water absorption and storage has evolved since the late nineteenth century. While LA water managers have long seen the northeastern Valley as potential infrastructural nature, the realization of those visions has largely been stymied by competing uses for and claims on the land with more direct links to processes of capital accumulation. The contemporary iteration, articulated through plans like the SCMP, aims to circumvent such conflicts by adopting a more spatially distributed approach to stormwater capture that largely preserves existing land uses and property arrangements. In other words, it is an attempt to restore pre-development hydrological functions here without threatening any of the development. But as I show, that project must still contend with the legacies—both material and social—of the racialized industrial development that has concentrated ecological hazards within this terrain. Sketching this arc, I explore how this approach stands to rework both water agencies' presence within the northeastern Valley and the landscape's role within LA's urban fabric, drawing new (not necessarily sought-after) forms of attention, investment, and labor into intimate corners of these neighborhoods.

Attending to these emergent forms of infrastructural nature underlines the spatially concentrated impacts of efforts to replumb LA. Presented by water managers and environmentalists alike as a transition that will increase citywide resilience to future droughts, this instantiation clearly directs most attention to a small section of the urban fabric—the porous terrain overlaying the groundwater basin—for this purpose. As such, the approach shares a spatial logic with initiatives that seek to mobilize carefully managed (and in some cases, strategically abandoned) lands to mitigate urban flood, landslide, or fire risks.[6] In such arrangements, selected landscapes are understood to buffer others from the disruptions associated with a changing climate. Conceptualizing the local iteration of these buffering functions as recharge work that the state seeks to draw from the landscape reveals the new relations and obligations, in addition to the much more explored risks, that such resilience-oriented projects can bring to sections of a city.[7]

STORMWATER-SPREADING GROUNDS
AS VESTIGIAL INFRASTRUCTURES

A few weeks before the public release of the draft SCMP, I met to talk stormwater with Ken, an engineering consultant the LADWP had hired to work on the plan. An expansive storyteller, Ken seemed eager to recount how he came to appreciate the resource potential of LA's runoff.

During the early months of 2005, he told me, he would spend his lunch hour down by the LA River, admiring how well its concrete-lined channel contained the city's storm flows. It was a wet winter in Southern California, and rain was often falling when he made these trips. "Just crazy, crazy amounts of water," he recalled, expressing awe at both the sheer volume of liquid sluicing through the channel and the infrastructures that prevented flood damage from the torrent.

Like the engineers who praised the technical ingenuity and vision required to build the LA Aqueduct, Ken was waxing poetic about a piece of Southern California's infrastructural skein that frequently draws criticism. Scholars and environmentalists alike have derided the region's extensive network of hardscaped storm drains, developed over the course of the twentieth century, as a short-sighted, ecologically destructive, single-purpose approach to managing runoff.[8] Embedded within such accounts is typically the contention that the network treats these flows only as a hazard, siphoning them out to sea at the expense of local aquifer recharge. Defenders of the concrete-lined waterways often emphasize the loss of life and property the storm drains prevent, suggesting that ungrateful contemporary residents take this protection for granted.[9] And some, like Ken, will also highlight the fact that managing runoff as a water supply resource is not an entirely new project for Southern California water managers; it's a goal built into LA's flood control system, albeit far less prominently than some of his predecessors envisioned.

Making this point in our interview, Ken told me the story of his first visit to the LADWP's Tujunga Spreading Grounds, a trip that took place soon after his lunchtime reveries by the storm-swollen river. Though he had lived in Southern California for years, he had never bothered to stop by (or really given much thought to) the northeastern Valley's five dedicated water-spreading facilities.[10] This is understandable: while not invisible within the urban fabric, the spreading grounds are easy to look past. The 150-acre Tujunga facility is located right where the Hollywood Freeway meets Interstate 5 near the northeastern edge of LA, about twenty miles from the downtown core. An attentive observer passing through this elevated intersection might notice an array of wide, shallow earthen and gravel basins arranged around the converging freeways, land fenced off from the surrounding streets. Though carefully designed to maximize

Figure 11. LADWP worker standing at the edge of the Tujunga
Spreading Grounds. Photo by author.

their water absorption function, the grounds resemble nothing so much
as wide, shallow pits full of furrowed dirt and gravel, connected by a few
pipes. The leaves of some scattered plants poke up through the rocks, but
there are no prettying flourishes. Unlike the US West's towering dams or
the thick aqueduct pipes that snake across the landscape along California's North-South freeways, spreading grounds do not particularly an-
nounce themselves as important water provision assemblages.[11] They look
like fallow fields, or spaces where someone meant to build something but
then forgot to follow through.

When viewed after a storm, however, the infrastructural character of
the grounds reveals itself somewhat more readily. An upstream dam re-
leases water into a concrete flood control channel that runs alongside the
grounds. Engineers open and close intakes within the facility, starting
and stopping the flow of the runoff from the channel into the basins. A
time-lapse video shot from above would capture the repetitive tempo of
the basins absorbing the water and the engineers refilling them, trying to
maximize aquifer recharge while minimizing local flood hazards.

Recounting his tour of the facility, Ken recalled his reaction: "Ohhh *this*
is how you get water into the ground!" He continued: "You need water and

good ground, and where those two things come together, that's where the greatest opportunities are. That's why they [the spreading grounds] are where they are. The Tujunga Wash ripping down the mountains deposited deep beds of gravel and sand, a perfect bed of sand and that's a perfect aquifer to recharge up there." The LADWP's hydrological models confirm that the recharge facilities contribute to the city's water supply, indicating that, on average, the Valley's spreading grounds augment the San Fernando Groundwater Basin with about twenty-five thousand acre-feet of stormwater annually.[12] Based on current LA consumption rates, this volume of water is enough to supply roughly fifty thousand households with tap water for a year. "That's significant!" Ken intoned after explaining the figures. "We can increase that by a significant amount, but you shouldn't say we're not doing anything."

Like Ken, I see the northeastern Valley's stormwater spreading grounds as revealing, often underappreciated sites within LA's water provision system. Though relatively marginal contributors to the city's water supply mix in volumetric terms, the grounds represent an old, half-realized managerial vision for the northeastern Valley's landscape and local water flows, one based on setting aside large tracts of land for the sole purpose of aquifer augmentation. Tracking the frustrated rise of this approach through the first half of the twentieth century helps to contextualize contemporary efforts to transform the area into a more diffuse sort of stormwater-capture landscape. Elaborating this very brief "shadow history" of the expansive version of the spreading grounds outlined in water managers' plans of that era offers a window into the multiple, conflicting versions of the northeastern Valley's future that once jostled for realization.[13]

Around the turn of the twentieth century, engineers were beginning to experiment with techniques to enhance the water absorption functions of the rocky lands near the foot of the mountains bordering much of LA County.[14] The northeastern Valley was considered a particularly compelling target for such initiatives. Speaking at the 1899 Annual Convention of the American Forestry Association, LA City Engineer Frank Olmstead made the case in enthusiastic terms. "It seems to me that there is a possibility of natural storage at the mouths of the canons [sic] by the diversion of the streams during the flood season over large beds of sand and gravel which lie there, that will materially increase the flow of the local

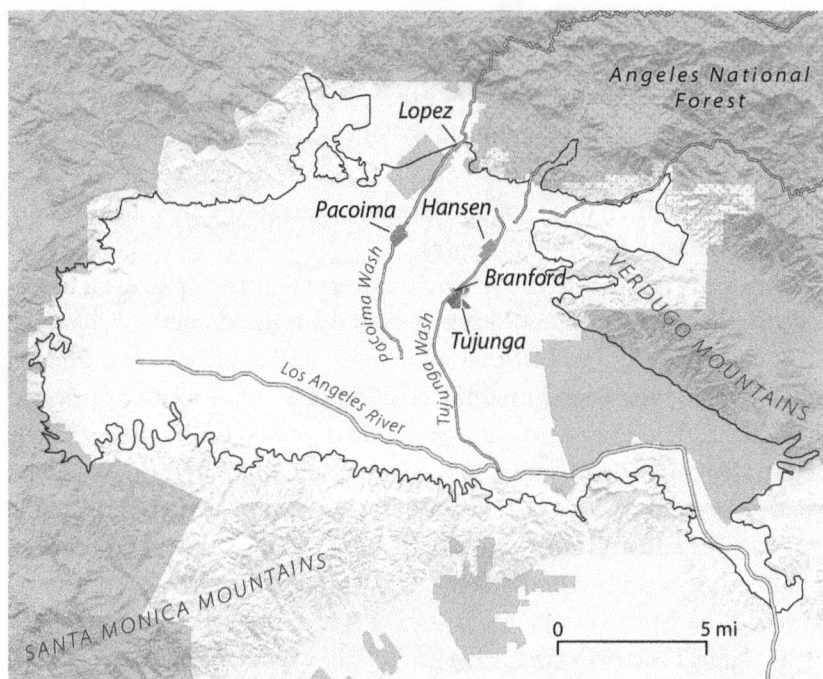

Map 4. Five stormwater-spreading-grounds facilities located in the San Fernando Valley. Though early twentieth-century engineers envisioned setting aside large swaths of the San Fernando Valley for stormwater recharge, today these installations only cover about five hundred square acres. Map by Nick O'Gara.

rivers," he told the crowd, emphasizing the hydrological connection between surface and subterranean waters. He continued: "In ordinary seasons the gravel absorbs all the flow, but when there is a heavy rainfall, a great part passes off to the sea, or on to lands that do not need the water. There are 12,000 acres of sand and gravel at the mouths of Pacoima and Tejunga [*sic*] washes. . . . Dikes should be built and the storm water in years of heavy rainfall distributed by means of strands, which will guarantee better absorption."[15]

Here, Olmstead articulates a simple infrastructural concept: let the gravel beds at the terminus of two LA River tributaries (the Tujunga and Pacoima Washes) keep absorbing the storm flows that thunder down from the mountains after a rain—as they already do. Just build some barriers around them so that the water stays there longer, allowing more runoff to infiltrate when there's a big storm. In effect, he advocates for setting aside a sizeable chunk of the Valley to preserve and subtly intensify the hydrologic function of the floodplain, a slight revision to more typical dam-and-surface-reservoir arrangements. In this context, recurrent inundation becomes a condition to be managed for, not against—at the expense of developing the land for other purposes.

Olmstead's queasiness at the image of stream flow that "passes off to sea" is echoed in the Flood Control Act of 1915, the legislation establishing the LA County Flood Control District.[16] Engineers and politicians in the region had for some time encouraged the creation of a public agency dedicated to managing flood risk across the county, and a ruinous 1914 deluge finally helped them marshal the necessary political will. But the text of the Flood Control Act suggests that the threats of flood and destruction were not the only concerns driving the creation of this institution. Sandwiched between an intention to control floodwaters and to protect property from damage, the enabling legislation presents stormwater conservation as one of the Flood Control District's organizing goals.[17] The agency's long-delayed Comprehensive Plan for Flood Control and Conservation reiterated this orientation. Adopted by the LA County Board of Supervisors in September 1931, the plan sketches a long-term vision for the development of stormwater infrastructures across the county. A newspaper account of a presentation by the district's chief engineer E. C. Eaton to the Board of Supervisors suggests the centrality of water salvage to the plan, highlighting its directive for the county to purchase ten thousand acres of land for percolation facilities.[18]

This didn't come to pass. Today, Flood Control (now part of the LA County Department of Public Works) owns and operates twenty-four hundred acres of spreading grounds across the county, less than a quarter of the area identified in their 1931 plan—and far, far less than the twelve thousand acres Olmstead envisioned setting aside within the San Fernando Valley alone. LA's infamous New Year's Day Flood of 1934 played an undeniable role in setting this trajectory. When the city received 7.4 inches of rain in twenty-four hours, the limited flood control infrastructure that the Flood Control District had managed to develop proved no match for the storm's runoff. Stormwater and debris filled the floodplain, destroying hundreds of homes and killing dozens of people. In the months that

followed, the agency's past efforts were widely deemed a failure, as the immense scale of damage from the storm undermined faith in its capacity to
manage runoff. In the popular imagination, the dangers associated with
LA's unpredictable rain now seemed far more palpable than its potential
to be efficiently corralled as a water supply resource. This outlook shaped
the local politics around stormwater management in the years that followed, further tipping the balance of the district's work toward the pursuit
of prominent, highly engineered flood control infrastructures like dams
and storm drains.[19]

Two additional, interconnected factors also helped circumscribe the
development of stormwater-spreading facilities, particularly within the
northeastern San Fernando Valley. As the careful reader will recall, Owens
Valley water began flowing into LA in 1913, two years before the Flood
Control District's establishment. This influx of Sierra Nevada snowmelt
dramatically reduced the city's reliance on local groundwater and surface
streams for its supply. The pipeline's enabling legislation forbade the sale of
its water to customers beyond city limits, a prohibition that directly precip-
itated the 1915 annexation of 170 square miles of the San Fernando Valley
into LA.[20] This move famously enriched land speculators who had bought
up swaths of the Valley shortly before the public announcement of the
aqueduct project.[21] The subsequent growth in property values discouraged
public acquisition of large tracts of northeastern Valley land for stormwater
percolation, leading the Flood Control District to purchase a relatively
modest 350 acres within the Valley for this purpose over the next four decades. Meanwhile, the presence of plentiful imported water effectively dis-
incentivized LADWP investment in stormwater capture. The presence of
that supply did contribute to the agency's decision to develop the Tujunga
Spreading Grounds in the early 1930s, but for the purpose of recharging the
San Fernando Basin with "excess" LA Aqueduct water, not local runoff.[22]
The latter capacity had to be built into the facility decades later.

Structured differently, New Deal–era legislation might have enabled
the Flood Control District to realize the Comprehensive Plan's ambitious
targets for stormwater salvage across LA County. But while the federal
Flood Control Act of 1936 dramatically increased federal funding for
stormwater management, it ultimately did so in a manner that encouraged the development of infrastructures with a different spatial logic.

Unlike 1933's Tennessee Valley Authority Act, which created a new federal agency organized around the conservationist ideal of comprehensive river basin development, the Flood Control Act laid out a limited purview for federal intervention in riparian systems.[23] While it established the federal government's role for flood management on all navigable streams and allocated $310 million toward this goal, the act parceled out responsibilities in a manner that limited the central government's role in this process.[24] Under the legislation, the Army Corps of Engineers could build large-scale, downstream flood control infrastructures, and various US Department of Agriculture agencies could construct small reservoirs near river headwaters. But local public agencies were stuck with the task of buying up land and easements, and water conservation projects were not considered federally fundable. Given LA's increasingly hot land market, this arrangement led the county Flood Control District to prioritize flood control arrangements requiring the least possible surface area.

The persistence of such understandings of urban land is apparent when contemporary water managers celebrate the groundwater basin as a water storage space desirable largely because it allows the ground above it to remain usable for other functions. In contrast to the Owens Valley, where in the early twentieth century LA acquired over 250,000 acres to manage for the purposes of maximizing the water available to transfer through its aqueduct (severely constraining local land development and dispossessing Native communities in the process), the city did not set aside these more proximate lands for the same purpose. A short news item from 1968 offers a useful snapshot of the understandings of urban land and water supply arrangements that structured this approach. As the (un-bylined) *LA Times* reporter explains, geologist Arthur Court was beginning to think that the spreading grounds might be outmoded. A professor at Valley State College, Court assigned a trio of undergraduates the task of studying the northeastern Valley's Pacoima Spreading Grounds, located near the school. The students analyzed the grounds in economic and hydrological terms, then delivered a damning verdict: "The operation of this spreading ground is not in the public interest."[25]

The students reached this conclusion through a simple analysis of property values. LA County bought the 175-acre tract of land for $40,000 in 1932 and listed its present-day value at $150,000, reflecting a 4 percent annual increase in value. The students, however, were skeptical, noting that the assessments of nearby properties "indicate the land may be worth at least $25,000 an acre for residential use or $60,000 an acre for commercial use, for a total value between $4 to $11 million."[26] Further, they noted that the grounds' utility as water supply augmentation infrastructure varied wildly, from a low of 78 acre-feet per rainy season to a high of 2,000 acre-feet over the past fourteen years. The risk of low yields was presented as a reason to halt the use of the land for water supply enhancement and release it to the market to encourage greater tax revenues. Speaking to the reporter, Court said that the study made him wonder: "Are we not behind the times in our water technology? At all events, no one can object to the conclusion of the students that a careful reexamination of present obsolete practices is needed." Built in an era characterized by lower land values and a more limited supply of imported water (before the Metropolitan's infusion of Colorado River water into the region), the grounds appeared to him incongruous—like vestigial traces of an outmoded resource management paradigm.

As Ken's delight at encountering the spreading grounds several decades later suggests, contemporary water managers see these facilities somewhat differently. And this has been an assessment backed by public resources: the LADWP has sunk tens of millions of dollars into upgrading the city's spreading facilities over the past two decades.[27] Even so, Court and his students' assessment is helpful in highlighting the dramatic transformation of northeastern Valley's land (and its attendant valuation) in the years following Olmstead's suggestion to set aside twelve thousand acres of the area for stormwater salvage. As the next section details, these decades of development fundamentally reworked the conditions of possibility for managing this terrain as water-holding infrastructure.

THE PROBLEMS OF A POROUS STORAGE TANK

Throughout my months of fieldwork, interlocutors used several analogies to describe the San Fernando Groundwater Basin. Likening the aquifer to a storage tank or an underground reservoir was the most common. Similar, though just a bit cuter, was the label "God's cistern," which I heard from a former environmental NGO staffer. All these characterizations, I thought, reflected LA's legal approach to the basin's water storage capacity. Though engineers' most ambitious, land-intensive stormwater capture visions for the northeastern Valley were ultimately not realized, the city poured substantial resources into securing its legal rights to water held in the aquifer below it. The formal process of adjudicating the San Fernando Basin began in 1955, due to disputes between LA and the neighboring cities of Burbank, Glendale, and San Fernando. Following a detailed 1962 study of local hydrological conditions, a trial court decision, and a judgment from the California Supreme Court in 1975, in 1979 Judge Harry Hupp issued the final ruling: LA was rewarded an undisputed priority claim, known as a Pueblo Water Right, to all the surface and subsurface water within the San Fernando Basin.[28] Though ordered to split "return water," the portion of imported water served within the basin that percolates back into the aquifer, with the other jurisdictions, the decision affirmed LA's right to all purified sewage and captured stormwater held within the basin.[29] In other words, this is not just some general storage tank; it's one to which the city has a special, protected claim.[30]

But my favorite nickname was a formulation that highlighted a characteristic of the basin beyond its capacity to stockpile water for future use. Speaking at a summer 2015 RWAG meeting, Oscar, a high-level manager at the Bureau of Sanitation, earnestly compared the basin to a bank, the water it held to money, and the pumps used by the city to extract that liquid to an ATM machine—that is, a broken ATM machine. "We're working to clean up the groundwater," he explained, describing a pilot treatment plant operating at that time near the Tujunga Spreading Grounds. "Think of it as fixing our ATM." Drawing such a comparison, Oscar was acknowledging a reality of the basin that is rarely highlighted in discussions of its stormwater capture potential: the groundwater that it holds is tainted by dangerous subsurface plumes of pollution.

The aquifer's condition is due to a confluence of uneven urban development and the intrinsic permeability of the northeastern Valley's soils. As the city's early efforts to manage the landscape as infrastructural nature receded during the middle decades of the twentieth century, the area emerged as a racialized industrial landscape, an area where communities and land uses undesired (though relied on) by whiter, wealthier sections

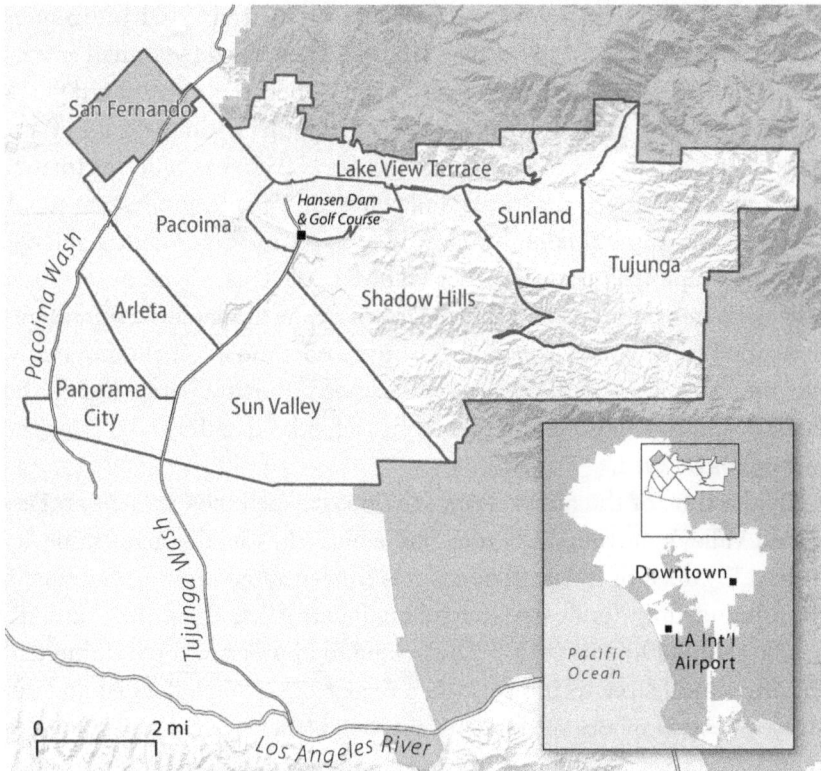

Map 5. Detail of the northeastern Valley neighborhoods prioritized for stormwater recharge projects. Map by Nick O'Gara.

of the city could establish themselves. As such, much like the hinterland areas that provide electricity or water for the city, the northeastern Valley has served a constitutive function for LA.[31] This development trajectory has had substantive material consequences, enabling the accumulation of groundwater pollution and flooding hazards within these neighborhoods. And as the reference to the city's need to fix "our ATM" attests, the spatial legacies of this period complicate contemporary efforts to shore up the LA water grid's reliability under future dry spells by stockpiling a subsurface water supply.

At the turn of the twentieth century, the entire San Fernando Valley was understood as LA's rural hinterland. By that point the Mexican cattle barons who had wrested control of Valley lands from local Indigenous communities had themselves been largely dispossessed by a small coterie of Anglo businessmen. Though the city and its boosters encouraged farming in the Valley, struggles to procure sufficient irrigation water initially limited the expansion of cultivated acreage.[32] The feasibility of further development changed drastically following the Valley's annexation to LA and the resulting influx of Owens Valley water in 1915. Planted acreage expanded rapidly in the years that followed, as citrus, olives, grapes, and other sun-hungry crops flourished. Drawn by the available water and undeveloped land, large-scale manufacturing operations—including many, like the giant Lockheed-Vega (now Lockheed Martin), connected to the defense industry—also began to establish a foothold in the Valley from the mid-1920s onward.

By the time of the annexation, the flood-prone lands of the San Fernando Valley's northeastern edge had emerged as its de facto "minority district."[33] Much of this settlement was concentrated around Pacoima, a town founded by Tennessee transplant Jouett Allen along the Southern Pacific railroad line in 1887.[34] Allen came to appreciate the flashiness of the area's local streams in 1891, when the Pacoima Wash flooded and destroyed several of the settlement's original structures.[35] Conceived as a suburban retreat for the wealthy, the combination of this periodic flooding and the real estate bust of the early 1890s set the area on a different path. As well-to-do whites departed, Black, Chicano, and Japanese American railroad workers and farm laborers settled along the railroad tracks, where they were able to purchase small plots of land.[36] Lacking the restrictive housing covenants that prohibited nonwhites from buying or renting in many sections of the city, the area developed as a peripheral, multiethnic hub, home to substantial Black, Latinx, and Japanese American populations by the middle of the twentieth century.[37]

The post–World War II period was an era of rapid population growth and land use transition across much of the Valley, including its northeastern corner. The defense industry's expansion during the war years had drawn a multiracial influx of working-class laborers to the area, presaging the uneven forms of suburban and industrial development in the decades that followed.[38] Notably, the extension of networked infrastructures, such as paved roads, streetlights, drainage facilities, and sewer hookups,

lagged throughout the Valley during the midcentury boom years.[39] Insufficient sanitary infrastructure proved particularly vexing. The feverish
pace of development caused a sharp increase in the volume of waste moving through the system, overwhelming its capacity by the early 1950s and
resulting in localized sewage spills in Valley streets.[40] The network's immediate volumetric difficulties were eased with the completion in 1955 of
the La Cienega-San Fernando Valley Relief Sewer. But continued growth
meant that there were frequent delays in getting new development connected to the system, leading many subdivisions and businesses to rely on
cesspools for waste disposal for protracted stretches.

The lack of amenities in and around multiracial Pacoima was especially pronounced. "Pacoima streets were unpaved, full of holes and rocks.
During a rain the rich, brown mud clung to our shoes," wrote Mary Helen
Ponce in her memoir of her midcentury childhood in the neighborhood,
noting the absence of indoor plumbing and electricity in many local
homes during the period.[41] Reverend Hillary Broadous, who moved to
the neighborhood in 1946, recalled similar conditions in a 1977 interview:
"Pacoima had one paved street that was Van Nuys Boulevard when I came
in 1940s. It had one paved street. There was no sewers, there was no sidewalk. I, along with others, walked the street, and got a petition for street
lights and sidewalks . . . and there was no mail service, that kind of thing.
Just the phone company" (Broadous 1977). Stormwater drainage also remained an issue. While several local dams, including the Army Corps'
massive Hansen Dam, were constructed to manage runoff flows, storm
drain construction was sluggish in the area. Exacerbated by the introduction across the Valley of impervious surfaces that limited the rainwater
absorption into the landscape, local flooding persisted during storms in
many sections of the northeastern Valley.

Such floods even roused the local dead in at least one circumstance, an
extraordinary happening I learned about from Nora, a middle-aged, life-
long resident of the northeastern Valley and a participant in a pilot stormwater capture project. Wedged into her living room couch on a Saturday
morning, she began her account of flooding in the neighborhood the way
that many did, with an emphasis on how storms would complicate efforts
to perform basic local errands. "Coming down Lanark [a local thorough-
fare], you would try to drive on it, and you're floating. I remember we

were like an hour and half late for school because the train underpass on Woodman had flooded," she recalled. Then she got to the surreal part. "The cemetery flooded out," she told me with a grimace. "People woke up with caskets in their yards, bodies kind of loose." Newspaper coverage from February 1978 corroborates her story, documenting thirty corpses wrenched from the grounds of the nearby Verdugo Cemetery.[42] As such events attest, while mostly experienced as a quotidian nuisance, local flooding has remained a hazard associated with the possibility of more serious (and sometimes, macabre) danger for northeastern Valley residents.

Despite these infrastructural limitations, during this period the area grew quickly and took on a strikingly industrial character, distinct in form, extent, and density from the manufacturing hubs in other sections of the Valley. Enabled by permissive zoning laws, much of this development took hold in and around the gravel mines that had since the early twentieth century marked the neighborhood now known as Sun Valley.[43] Once retired, many of those facilities were converted into landfills, and a range of junkyards and auto wrecking lots were established nearby.[44] Machinery and metallurgy operations also clustered in the northeastern Valley.[45] As others have shown, the concentration of nonwhite Angelenos and polluting industries in these neighborhoods was not coincidental but was produced and sustained through local land use planning processes and racially exclusionary housing and transportation policies.[46] Sited within a landscape crisscrossed by several freeways, the northeastern Valley's sprawling industrial cluster contributed to the degradation of local air and soils, even decades after many of the facilities shuttered and the associated employment opportunities disappeared.[47]

The effects of the Valley's uneven midcentury industrialization on the San Fernando Groundwater Basin became a formal object of concern for LA following a 1979 directive from the California Department of Health Services requiring that water purveyors test groundwater for industrial chemicals. In the northeastern Valley, samples revealed concerning levels of two common solvents, trichloroethylene (TCE) and perchloroethlyene (PCE), in local groundwater wells.[48] Further testing indicated that the pollution was particularly widespread in eastern sections of the basin, due to the confluence of industrial waste disposal practices and the permeability of local soils. Firms across the Valley had been directing their waste

Figure 12. Typical northeastern Valley industrial thoroughfare.
Photo by author.

to cesspools and leach fields for decades, due in part to the hassle and
expense of connecting to the city's underdeveloped sewer network.[49] Beyond cost concerns, the approach was driven by the assumption that the
landscape could provide a particular sort of cleansing work—that is, as
LADWP engineer Larry McReynolds put it, that "the ground would purify
just about anything."[50] Instead, the porous soils on the eastern side of the
Valley allowed the hazardous substances to spread. "We are more concerned about the East Valley because it is a much more open aquifer—full
of sands and gravel and things like that," McReynolds explained to the
LA Times during the early phases of the investigation.[51]

The absorbent soils looked even more dangerous after a 1985 study revealed that thirty-seven hundred industrial and commercial businesses
and apartment complexes in the Valley still relied on septic tanks or cesspools for waste disposal.[52] Subsequent tainted samples revealed growing
plumes of TCE and PCE as well as the presence of other pollutants, forcing the LADWP to remove dozens of water supply wells from operation.
The list of "broken ATMs" has only grown since the 1986 classification of
a large section of the basin as a federal Superfund site, further limiting
the city's draw from its subsurface water source. After decades of basin

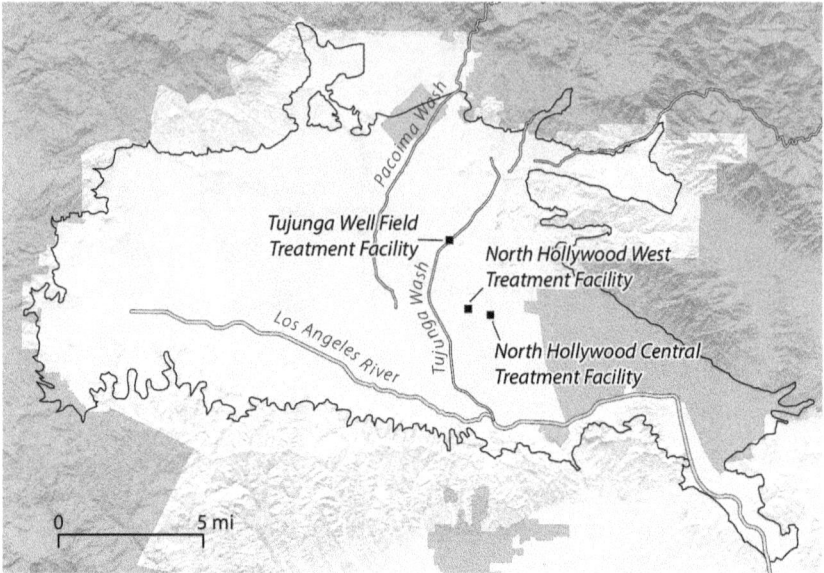

Map 6. LADWP groundwater cleanup facilities in the San Fernando Valley, developed more than forty years after extensive plumes of pollution were discovered within the San Fernando Basin. Map by Nick O'Gara.

characterization studies and legal wrangling, at the time of this writing a long-term solution appears to have finally arrived, as the LADWP is scheduled to bring a $587 million trio of tax-, ratepayer-, and polluter-funded groundwater treatment plants online by late 2024.[53] Beyond the slow pace and staggering cost of the cleanup, the trajectory is also notable for how it adds another, less salutary valence to the notion of landscape permeability, as presented in contemporary stormwater plans. The northeastern Valley's alluvial soils, it turns out, can enable not only planned-for water absorption, but also the migration and concentration of toxic pollution. The distinctly relational materiality of groundwater thus means that porosity can carry danger in addition to opportunity.[54]

As these histories attest, contemporary projects of infrastructural nature in the northeastern Valley seek to rationalize a landscape that has been deeply marked by a long-term concentration of localized environmental hazards. Nora's image of caskets swept from cemetery grounds by floodwaters and deposited onto front lawns is both unsettling and helpful in conveying the sense of danger and disarray that many of its residents associate with the resulting terrain. Persistent infrastructural neglect, rather than the attentive, resource-maximizing management that recent stormwater-capture proposals suggest, has long characterized the state's approach to this section of LA. Tracking the northeastern Valley's development as a racialized landscape of ecological disorder thus clarifies a notable dynamic that underpins the stormwater-capture projects detailed in the next section: those efforts seek to cultivate ecological services *for* the city from a landscape where marginalized communities have long received inadequate public services *from* the city.

SEEING A LANDSCAPE OF DISTRIBUTED STORMWATER POTENTIAL

Seated at a conference table in the LADWP's downtown headquarters, Norm, an early-career engineer for the agency, told me that he'd been wildly busy throughout the drought-time spring of 2015. He and his colleagues had been hard at work analyzing a handful of sites in the northeastern Valley where the agency might develop stormwater infiltration projects in the coming months. He spoke at some length about one promising but challenging installation, slated for construction within a grassy power-line easement. The LADWP owned the land, a site with good hydrological connectivity with the aquifer below. And its flow models suggested that more than one hundred acres of the surrounding landscape drained to the easement, making it a strong candidate for high-volume stormwater capture infrastructure. The agency wanted to build some infiltration basins on the site to direct that runoff to the aquifer, but the underground utilities were complicating the process. The city's Bureau of Engineering would eventually take over the finer points of the design, but Norm needed to provide a defensible stormwater infiltration estimate for a feasible project, not an easy task in this case. Even so, he seemed happy to be working on the site. "This got the highest priority of the stormwater projects," he told me, referring to an analysis in the draft SCMP. "It's exciting to see that we're moving it forward."

This focused approach to the northeastern Valley's urban runoff represents a marked shift in local water agencies'—particularly the LADWP's—relationship to this section of the city. When I asked Norm's boss Luis to describe the team's work, he made it clear that their intention was to install infiltration projects across the landscape: "We're scouring the northeast Valley, looking at these flood maps, going section to section. We want to get good runoffs there, get the highest volumes, then the next and the next, blanket the whole area." The details of the SCMP reflect this geographic focus, as the first seventeen "centralized" stormwater projects it calls for were all slated for (or underway in, or recently completed in) this section of the Valley.[55] And in recent years, such projects have begun to attract substantial city investment. Between 2007 and 2017, LA spent $120 million on stormwater recharge projects in the landscape above the San Fernando Groundwater Basin, include major spreading-grounds upgrades.[56] The city's 2019 Green New Deal set targets of adding ten new stormwater capture projects by 2025 and one hundred by 2035, highlighting the anticipated acceleration of such work. If the infrastructural development outlined in these plans is carried out, within a few decades stormwater will move through this space far differently than it did the day I spoke with Norm in 2015, signaling an attendant shift in water agency priorities and practices.

Within LA, many of my interlocutors, both water managers and environmental NGO workers, trace the rise of such efforts to retrofit groundwater recharge capacity into the lands of the northeastern Valley back to the work of the late environmental advocate Dorothy Green. By the end of the 1980s, Green had become an influential local activist through her work organizing against a large-scale water transference project and for a higher treatment standard at Hyperion. These fights behind her, Green sought a project that could address all the city's big water issues—supply, flood control, and water quality in local streams—at once. She landed on the notion of unpaving LA for the sake of soaking up more urban runoff. As Green told oral historian Susan Collings in a 2006 interview:

> It occurred to me that if we're ever going to take any concrete out of the rivers and streams that could deal with improving water quality, because vegetation really helps a lot, to improve water quality, that we would have to manage storm water in a better way than what we're doing. And clearly, one

way to do that is to unpave as much of the city as we can, to get that stormwater into the ground. It would improve quality of life in the city, as well as
augment our drinking water supply, as well as reduce what's in the storm
drain system, so that maybe we can take—restore some of those creeks and
rivers. Just the multiple benefits were really clear early on.[57]

While the language of unpaving failed to gain traction, the concept of
retrofitting lost absorbent capacity back into the urban landscape was
taken up by a range of local environmental groups in the years that followed, threaded through campaigns and lawsuits related to flood control,
water quality, and water supply issues. Such sustained pressure was widely
credited with public agencies' growing interest in green infrastructure
projects, as signaled by documents like the SCMP.

Many of the proposed runoff capture projects—including Norm's infiltration basins—fall into the category of green infrastructure. Such installations don't look like what most people imagine when they hear the
word infrastructure. But like other infrastructures, these are designed to
perform a particular support function, directing the flow of resources in
a specific way. Typically, green infrastructure is meant to maximize the
landscape's ability to absorb water. For instance, an installation might involve the replacement of a concrete street median with a strip of vegetation, planted a few inches below the pavement around it, with breaks in a
curb to siphon stormwater in from the street. Carefully selected mulches,
soils, gravels, and plants will be arranged to absorb as much runoff water
as possible.

The US Environmental Protection Agency's (EPA's) definition of green
infrastructure, excerpted here, is revealing in its emphasis on the work it
ascribes to slices of land reconfigured as green infrastructure installations:
"Green infrastructure uses vegetation, soils, and other elements and practices to restore some of the natural processes required to manage water
and create healthier urban environments. . . . At the neighborhood or site
scale, stormwater management systems that mimic nature soak up and
store water."[58] In this rendering, the "natural processes" manage water and
create healthier environments, framing the functions performed by the
soils and plants in a manner that echoes ascendant discourses on the potential role of nature's or ecosystem services in environmental management. As the EPA's broad language—"manage water" and "create healthier

Figure 13. Slightly unkempt small-scale green infrastructure installation in the northeastern Valley. Photo by author.

urban environments"—suggests, green infrastructure projects are not always oriented toward aquifer recharge or water supply augmentation. Such installations' capacity to absorb runoff is also framed by proponents as a boon to overloaded urban drainage systems, an approach with the potential to address both water quality impairments and flood risk. With encouragement from the EPA, over the past two decades many US cities (including LA) have incorporated green infrastructure into attempts to remediate waterways classified as unacceptably polluted under the federal Clean Water Act.[59]

Notably, the EPA's formulation echoes former LA city engineer Frank Olmstead's idealized account of how the sand and gravel beds at the mouths of the Tujunga and Pacoima Washes might be mobilized, with the help of some strategically constructed dikes, to absorb larger volumes of water for the city's water supply. Both visions are premised on the notion that land can be designed and managed for the purpose of directing runoff to the aquifer. But the differences also deserve elaboration—in particular, the EPA's suggestion that such functionality can be achieved at the "site scale." Olmstead's approach was, after all, premised on the city's ability to dedicate twelve thousand northeastern Valley acres exclusively to runoff capture. While acknowledging the value of spatially extensive installations, the contemporary green infrastructure paradigm also asserts the efficacy of far smaller infrastructures capable of managing flows from surrounding terrain, like Norm's infiltration basins on the power-line easement. In contrast to approaches that would set aside large swaths of land for this purpose, such distributed green infrastructure programs aim to build water capture functions into the landscape without displacing established uses or ownership arrangements.

I heard a lot about the cost of land when I discussed these plans with water managers. In their telling, the cash-poor agencies that employ them are now tasked with addressing a complex set of urban water provision, water quality, and flood management problems—and all solutions carry massive price tags, to be borne by the public in some form. Kumal, an engineering consultant who had previously worked for the county's Flood Control District, made this point to me by recounting a report that an old boss had requested from him back in the late 1990s. Frustrated with local environmentalists' criticisms of LA's hardscaped storm drain system, the more senior engineer asked Kumal to calculate how much it would cost to rip out the system's concrete and "naturalize" the LA River and its tributaries. Based on his hydraulic modeling of local storm flows and the available land valuation data, Kumal concluded that such a program would cost a cool $340 billion, due largely to the cost of acquiring the necessary acreage. Even the much more modest goal of naturalizing a one-mile stretch of the northeastern Valley's Tujunga Wash penciled out to over $100 million, due to the assumed need to purchase channel-adjacent parcels to absorb overspill during storms. For Kumal, such projected costs signaled the irrelevance of demands or plans for spatially extensive green infrastructure installations within LA. In keeping with this orientation, when we met he was consulting on a project that would install about a dozen infiltration basins into the narrow strip of grass between the street and the sidewalk—a public right of way with zero associated acquisition costs for the public agencies involved.[60]

Similar spatial logics and property arrangements marked many of the recharge projects under development or discussion in the northeastern Valley during my fieldwork period. I spent the most time observing and assisting around the neighborhood retrofit pilot project that Ellen, a key environmental NGO interlocutor, oversaw. With funding from several sources, including the LADWP, her organization had partnered with other local NGOs and green contractors to develop rain gardens, parkway swales, and other runoff-capturing infrastructures onto participants' properties. But while I knew it best, Ellen's was just one of several such micro-scale projects underway in the northeastern Valley during those years. Another NGO was helping to maintain a different effort just a few miles away, and several of my water agency interlocutors were involved in the early-stage outreach work on a similar project in a nearby northeastern Valley neighborhood.

Local water managers' growing familiarity with this scale of stormwater-capture infrastructure was reiterated for me at an EPA-sponsored green infrastructure design charrette in the fall of 2015.[61] This daylong event was held at the recently constructed environmental education building within the Hyperion facility's sprawling footprint. After a morning of familiar speeches and a catered lunch, I found myself sharing a table with Luis, Ellen, and some flood control engineers and staring at a detailed street map of a northeastern Valley neighborhood. Our group's task was to mark up the map with the new water infrastructures we would like to see built, then to share our vision with the forty or so EPA, local agency, and NGO workers gathered for the event. Luis took the lead, drawing infiltration swales and drywells on the narrow parkway strips between the sidewalk and the road. He talked a blue streak about the vagaries of overhead and underground utilities and how the public got excited about infiltration basins because they're pretty, but he really preferred the dry wells, because they accumulate debris more slowly. He had clearly gotten used to imagining the northeastern Valley dotted with aquifer-augmenting green infrastructure and envisioning his agency's role in this transformation, attending to tiny strips of public land and even the possibility of drawing in the capacity of engaged residents' suburban gardens.

While most of my conversations with water managers and environmental advocates framed these infrastructures as groundwater recharge projects that brought local environmental benefits to residents of the northeastern Valley, some would occasionally discuss them in terms that signaled a more instrumentalized view of the communities that would live within the new landscape of recharge. Hal, my interlocutor from the

RWAG who also frequently attended stormwater-focused stakeholder events, once invited me to lunch to discuss an idea he was very proud of: bucketless water buckets. By this phrase, it turned out, he simply meant compost, which he'd recently learned absorbs runoff far better than does hard dirt. While used to long conversations about the absorbent capacities of compost-rich soil by that point, I was taken aback by his proposed method for developing such ground in the northeastern Valley. The Bureau of Sanitation, he explained, already produces compost from yard waste and already owns trucks—so workers could just drive up and down the streets of Pacoima and Sun Valley dumping it on bare patches of people's front yards. When I ventured to ask whether he'd talked to anyone in those neighborhoods to gauge the level of interest in the idea of the city depositing piles of decomposed yard clippings in front of their houses, he admitted that he had not—but that he would expect them to welcome that sort of largesse, especially given how much it could help the city. Months after this discussion, Hal told me that city staff hadn't seemed interested in his concept, and their reactions mostly just left him disappointed. If only they were more open-minded, he lamented, they'd use all the tools at their disposal to solve the city's water problems.

Ideas like Hal's demonstrate how easily the desires and concerns of residents can recede when outsiders reimagine a landscape as terrain that might be managed to provide ecological functions. Within much of LA's stormwater discourse, the spatial arrangements necessary to absorbing and storing urban runoff draw more attention than their social and political ramifications. The people residing in the northeastern Valley and their concerns about and hopes for the local landscape have not driven these managerial visions. When it came to project delivery, the water managers and environmental advocates I spoke with were thoughtfully engaged with the process of community outreach and engagement. But this process never appeared smooth or seamless. As I detail further in the following chapters, my time in the northeastern Valley suggests some local ambivalence around the growing footprint of water-focused NGOs and public agencies within this landscape. Walking through an alleyway retrofitted for stormwater capture near a neighborhood recharge project one summer evening, I was struck by the concentration of graffiti on the sign announcing the recently completed project's name and funders. My

Figure 14. Graffiti-covered signage alongside a green infrastructure installation in the northeastern Valley. Photo by author.

interviews with NGO and city staff involved with such installations soon clarified that this sort of sign-tagging was common around these projects across the northeastern Valley. Though celebrated on NGO Twitter feeds and city planning documents as harbingers of a green, resilient future for the city's water supply system, these infrastructures often appeared somewhat less beloved by the communities living around them.

As this chapter has shown, many now understand and represent LA's northeastern San Fernando Valley as a landscape that could—and should—be managed to capture and store new water supplies for the city. In documents like the SCMP, quotidian spaces like suburban yards and parkway strips join parks and watercourses within the category of urban nature, while properly managed nature takes on an infrastructural role, due to its ability to absorb water. This is a vision that draws new lands and water flows into both the LADWP's gaze and the city's water provision system. In contrast to the twentieth-century engineers who built spreading grounds on northeastern Valley land dedicated exclusively to that function, contemporary water managers aspire to transform small slices of lived-in sections of the urban landscape into water-capture infrastructures.

While such aquifer-oriented aspirations for this land are clearly not new, trickles of public funding and plenty of mainstream environmentalist attention buoy the current iteration, a substantive departure from the forms of neglect that marked the land through much of the twentieth century. The LADWP's new, more engaged approach to this slice of the urban landscape is conditioned by its imagined future as a space with the potential to buffer the city's water network from anticipated shocks to other parts of its water provision system. While the agency is no stranger to considering the implications of land management for water extraction, until recently the LADWP reserved such attention largely for the Mono Basin and Owens Valley landscapes from which the agency draws its aqueduct water. After more than a century of looking ever farther afield for water supplies, the city is shifting its gaze inward to produce the vital resource—and to take on new projects of cultivating functions from slices of urban land.

This approach is grounded in an understanding of in-city terrain uncommon within urban social and spatial theory. For decades scholars have foregrounded cities' material links to their hinterlands (and in some formulations, to the entire planet through lengthy supply chains), demonstrating the importance of theorizing the production of urban environments as an ongoing process with stark implications for landscapes beyond the metropolitan boundaries.[62] In such formulations, the urban landscape emerges almost exclusively as a site of resource consumption. While not discounting such patterns—and indeed, keeping in mind that LA is pursuing this new in-city resource provision arrangement while fighting to sustain its claims on faraway waters—exploring the rise of this inward-looking paradigm demonstrates the growing importance of metropolitan terrain for critical resource *production*. That these alternative spatial logics are also producing new (and sometimes fraught) in-city extractive arrangements in the name of water provision and climate adaptation further signals their significance for urban spatial and social relations in the coming decades. The chapters that follow further elaborate the frictions, relations, forms of work, and processes of resource valuation emerging as LA pursues such recharge projects.

4 Infrastructural Work and Infiltrating Runoff

In February 2014, LA held a ribbon-cutting ceremony to celebrate the completion of the northeastern Valley's Woodman Avenue Multi-Beneficial Stormwater Capture and Median Retrofit Project. Nearly one hundred people attended the Saturday morning event, which was marked by speeches from representatives of the city's Bureau of Sanitation, a local NGO, and a congressman with ties to the neighborhood. Funded by a combination of money from the LADWP and Sanitation and a grant from a state bond measure, the $3.4 million project transformed a concrete median into twenty-seven thousand square feet of planted, terraformed berms and swales, designed to siphon in flows from the adjacent street via inlets cut into the curb. Developed to absorb runoff from over 120 square acres of an adjacent neighborhood, the installation was widely touted as a bold experiment in distributed stormwater infiltration to the city's reserve stockpile in the San Fernando Groundwater Basin.

But just a few weeks later, as water flowed in during a late-winter downpour, the infrastructure collapsed. In the city for a brief research trip in March 2014, I observed the aftermath firsthand with Hank. He was working on a house near the median that week and suggested stopping by the site to check it out on our way to lunch. When I commented that it looked much better than the mess of construction equipment and dust I remembered from the previous summer, he pointed out the sections of soil that had washed out with the rain. His theory was that the engineers designing the installation had underestimated the sharp grade of the median and constructed a truly precarious set of berms and swales. Ruminating on the failure over fish-ball noodles in a nearby strip-mall Thai restaurant, he wondered aloud how much of the touted 1.5 million gallons of runoff per storm the broken median would actually be absorbing before someone managed to repair it.

This encounter with the median stood in stark contrast to its tidy, rationalized appearance in the SCMP and associated promotional materials throughout my fieldwork months. There, Woodman surfaced as a pilot project connected to a very alluring number. Based on the LADWP's contribution to the project ($1.2 million) and the runoff that modelers predicted it would infiltrate to the groundwater basin over the course of the project's lifetime, the agency had priced each acre-foot of eventual drinking water from the median at $727, not dissimilar to the cost of imported Metropolitan water purchased by the city in the mid-2010s. The fact that much of the infrastructure had been damaged within weeks of installation (and that repairs were slow to follow) went unmentioned in this pitch for the SCMP and its cost justification. It also rarely featured in my NGO interlocutors' public comments on the median, which tended to emphasize the ingenuity and ecological promise inherent in the notion of replacing concrete with planted swales. Though not entirely fulfilling its intended stormwater infiltration function in practice, the median was useful for solidifying an account of infrastructural capacity and resource value understood as essential to replumbing the northeastern Valley into a landscape of groundwater recharge for the sake of climate adaptation.

This chapter examines green infrastructure advocates' work to stabilize such interpretations of distributed runoff-capture installations. Undertaken primarily by NGO staffers and younger engineers at public agencies, through these labors my interlocutors sought to advance the project of recharge, accelerating the movement of projects from the shiny pages of city plans to the soil of the northeastern Valley. To riff on anthropologist Nikhil Anand's much-cited articulation of how techno-political "pressure" operates within urban networks, here stormwater infiltration is realized by enabling understandings of green infrastructure's efficacy, monetary value, and fundability.[1]

Following recent scholarship across anthropology and geography, I approach these efforts as infrastructural in two senses of the word: they are connected to the construction of material infrastructures and they also underpin the realization of those projects.[2] Like the work of building and maintaining such installations, the technical and narrative practices I trace here are necessary for establishing and sustaining these facilities, and thus to drawing new stormwater flows into the city's water supply system. Put differently: this phatic labor is part of the process of making the northeastern Valley landscape's water infiltration services appropriable by the city.[3] Rendering the berms, rocks, and plants of installations like the Woodman Median legible across domains as technically efficacious, valuable, and fundable water-capture assemblages relies on strategic translational work, which involves but also exceeds the labor of environmental modeling.[4] Approaching this discursive infrastructural work as a key dimension of such stormwater assemblages thus helps to capture the full range of human exertion involved in realizing provisioning services from the urban landscape.

Somewhat to my surprise, I spent a considerable chunk of my fieldwork months observing these forms of work. And on a certain register, that reaction can be read as a sign of my interlocutors' success in disseminating their aspirational accounts of runoff capture in the city. Reading planning documents and newspaper coverage related to LA's stormwater ambitions from afar, I had the impression that the construction of infrastructural nature was proceeding rapidly and that I would spend my main research period observing fast-paced project installations. A lengthy opinion piece in the *New York Times* in December 2014, titled "Los Angeles, City of Water," is representative of such framings, suggesting that extensive replumbing was already well underway: "Recently, however, Los Angeles has reduced its reliance on outside sources of water. It has become, of all things, a leader in sustainable water management, a pioneer in big-city use of cost-effective, environmentally beneficial water conservation, collection and reuse technologies. Some combination of these techniques is the most plausible path to survival for all the cities of the water-depleted West" (Leslie 2014). But as situations tend to when viewed at a closer range, this one looked less coherent from the vantage of the NGO offices and water agency conference rooms where I spent much of my fieldwork time (not to mention from the sidewalks alongside recently constructed runoff-capture installations). The efficacy and value of, as well as the viable funding streams for, the northeastern Valley's anticipated array of green infrastructures all remained ambiguous in these spaces. By and large, what I encountered was not quite the uncertain interregnum following an infrastructure project's commencement and preceding a discernible endpoint that anthropologists have termed *suspension*.[5] Rather, it seemed like an awkward disjuncture between proudly trumpeted city intentions for rapid spatial transformation and actual moments of new project groundbreaking. Frustrated with these circumstances, which they viewed as a symptom of public agencies' maddening inertia, my interlocutors turned to technical and narrative tools in attempts to shift or stabilize key stakeholders' assessments of green infrastructure installations as effective and ultimately investable by the state.

Much of this work sought to establish new understandings of the material connections between runoff flows, green infrastructure installations, and the groundwater basin below.[6] In the face of unfamiliarity or skepticism, the NGO staffers and supportive water managers I observed were striving to produce a shared appreciation of the verticality of the northeastern Valley's watershed—as well as of their capacity to manipulate flows within that plane via the construction of distributed, nature-based facilities.[7] Complementary to recent scholarship detailing how ecological connections can be revealed through engagements with infrastructural nature projects, here I foreground the sustained human effort that often precedes the recognition of such relations.[8] While the liveliness of these soils and plants clearly matters, so does the labor of presenting and narrating those arrangements. Appreciating the active and intentional work that goes into engendering understandings of these connections reorients notions of novel infrastructures as "ontological experiments" that can help to seed the emergence of new entities and relations, enabling them to better account for the situated human agency and micro-social practices involved in such processes.[9]

Related efforts sought to establish firmer links between green infrastructure installations and official practices of ascribing monetary value to nature's functions. As I show, while drawing on hydrological models and established accounting techniques, the advocates' work aimed to spread more expansive, experimental notions of ecosystemic value, taking advantage of the unsettled state of practice around assigning price to nature's labor. Attending to these (frequently stymied) attempts to reorient established paradigms of knowing, valuing, and managing stormwater reveals the barriers—epistemological, metrological, and institutional—that can circumscribe projects of replumbing the city.

STABILIZING RECHARGE CAPACITY

"If you could just walk me through the calculations again?" Nick asked gently, his perplexed grimace oriented toward Ellen's large monitor screen and the adjacent speakerphone setup. Not for the first (or the last) time during my fieldwork, Nick and Ellen were sitting in the NGO's compact office on a call with LADWP's Luis and Norm to discuss stormwater modeling. That day the group was reviewing the engineers' current estimates of how much runoff was being directed to the San Fernando Groundwater Basin by the NGO's neighborhood retrofit project and discovering several points of disagreement. Listening to Norm review the math one more time, Nick identified the embedded assumption leading to the biggest discrepancy: the LADWP staffers believed that the green infrastructure installations' soil was absorbing water far more slowly than he and Ellen believed that it did, based on their extensive visits to the sites. A polite (if slightly exasperated) back and forth ensued, eventually leading the group to settle on a compromise runoff coefficient that better reflected the in situ observations. This was, I understood from my months around the office, a tweak that would improve the cost per acre foot of stormwater estimate associated with the project and bolster future iterations' chances at getting funded and built.

As such exchanges suggest, decades after ecosystem services' rise to policy prominence, the process of measuring and modeling the work of nature for the purpose of assigning it monetary value remains very much ongoing. Grounded in practices that aim to "translate parts of nature into calculable beings," these efforts rely heavily on ecological and hydrological sciences

Figure 15. Runoff pooling in a domestic-scale green infrastructure installation in the northeastern Valley. Photo by author.

to produce simplified representations of natural systems and enable investment in their continuing or enhanced functioning.[10] As others have demonstrated, the work of gathering, organizing, and interpreting the appropriate data for such purposes is a decidedly more than technical undertaking, often relying on idiosyncratic measurement practices and the situated value judgments of modelers.[11] Geographer Morgan Robertson's classic account of field ecologists struggling to identify indicator species that could help to make certain parcels eligible for classification as wetland mitigation sites is a particularly deft illustration of the exigencies of this work.[12] A key orienting factor in that context is the accepted goal "to produce data that successfully circulate in the networks of law and economics" through ecological fieldwork.[13] In a similar vein, creating a representation of a slice of infrastructural nature that will be useful for decision-making in other domains was the explicit purpose of Luis and Norm's green infrastructure modeling. But as Nick and Ellen's response to their draft model suggests, terms like *useful* are about more than just making the installations legible to new audiences. They can also make these retrofitted slices of nature emerge in those settings as credible, desirable stormwater conduits, advancing the project of developing more such facilities.

The importance of this credibility was understood among my NGO interlocutors as particularly vital in the case of distributed green infrastructure installations. Despite their prominence in the city planning documents, questions about these infrastructures' efficacy as groundwater recharge technologies never fully dissipated during my fieldwork period. This skepticism was especially common among the older LA County Department of Public Works engineers that I encountered. "When it's local, green streets or something, you don't actually know if anything gets to the groundwater table," Ned, a veteran of that agency, told me, one of several such asides littering my notes from conversations with his colleagues.[14] Recounting his years of working in spreading-grounds operations, Ned compared those facilities favorably with more distributed approaches to stormwater infiltration. "With centralized, you know it's getting down to the actual groundwater table," he concluded.

His doubts aligned with what geographer Mike Finewood terms a *grey epistemology*, a framework for understanding water "that focuses on the technical, abiotic aspects of a system" and favors the sort of centralized, concrete-heavy, runoff control infrastructures public agencies built out over the course of the twentieth century in LA, largely due to their readily calculable volumetric efficacy.[15] As Finewood details in his studies of green infrastructure planning in Pittsburgh, grey epistemologies can severely delimit the extent and form of nature-based installations considered or developed within a city.[16] I observed that many water managers, particularly those close to the end of their careers, shaped their work around similar paradigms of apprehending urban runoff.[17]

Some of the doubts about the installations' capacity to augment the city's water supply were connected to the invisibility of the flow path from the ground-level installation absorbing the runoff to the aquifer's subterranean crevices. Materially complex and infrequently visualized, aquifers remain mysterious and ambiguous environments or "inscrutable spaces" to many, including most water managers.[18] Despite the San Fernando Groundwater Basin's heavily litigated and polluted status, this characterization fit that space during my fieldwork period. Though the EPA and LADWP had collaborated on a 1992 remediation analysis of the basin to guide Superfund proceedings, relatively limited characterization data was gathered during the two decades of legal limbo that followed. Signaling a

new seriousness about pursuing large-scale groundwater cleanup and recharge, in 2015 the LADWP completed an $11 million study of the basin's
plumes of pollution. The agency drilled twenty-six new monitoring wells
as part of the research, which will be central to documenting the effects of
long-term remediation operations. As the ongoing need for such data suggests, many dimensions of this prospective water storage space, including
those related to the movement of substances within the underground formation, remain relatively opaque, even to local experts.

But uncertainty about the ability of green infrastructure to recharge
the basin was also grounded in concerns and knowledge gaps closer to the
surface—that is, regarding the capacity of such infrastructures to absorb
runoff flows in the first place. These worries were related to the material
and temporal qualities of both local stormwater and the terrain itself. As
previously noted, LA in general and the northeastern Valley in particular
are known for flashy storm events and swift movement of runoff, which
are seen as more likely to overwhelm small infrastructures than centralized facilities. The region's longer-term patterns of drought and downpour
also present challenges. As spreading-grounds critic Arthur Court's students documented back in 1968, the volume of water infiltrated by stormwater facilities large and small fluctuates dramatically year to year. Several
spreading-grounds operators that I interviewed took pains to emphasize
that this variation is rooted in the duration, frequency, and intensity of
rain events within a given wet season.

"I've seen storm seasons that have gotten almost no rain in, then lots
of rain," Charles, another county spreading-grounds operator explained.
"Based on how the storm season went, percolation rates at the grounds
change. Regardless of the maintenance we do, groundwater conditions have
an impact on how much of the water gets into the ground." Though younger
than Ned, Charles had been working with those facilities for a decade and
had accumulated plenty of twelve-hour shifts on-site at the grounds during
storm season in that time. He had, as he put it, both a technical and a practical understanding of how recharge worked. Glancing out the window into
the late fall haze, he noted that they could probably absorb a huge volume
of water right then, given the protracted drought—but that during the rainy
winter of 2004–2005, he'd watched the grounds get so saturated that recharge capacity declined substantially by the middle of the wet season.

Discussing his years of work at the spreading grounds, Charles raised another worrying point about the volume of infiltration that they facilitate: many of the long-term absorption figures are based on estimates. At the time of our interview, he explained, his agency had only recently begun to install sensors within the grounds to provide real-time monitoring data on the speed of percolation. As the novelty of this form of quantification suggests, the work of effectively measuring the water-capture functions of even centralized infrastructures remains incomplete. Unsurprisingly, these challenges of monitoring and legibility are even more pronounced for newer, smaller-scale, distributed runoff capture installations within the city.

James, a hydrological modeler for the county, helped me appreciate the assumptions involved when we spoke about his simulation near the end of 2015. At that point he had spent years refining a mass balance model to assess pollution urban runoff caused in area streams, a tool meant to help local jurisdictions comply with the federal Clean Water Act water quality standards. He explained that while he incorporated localized rainfall and stream gauge numbers into his simulation, there was virtually no infiltration data available outside the limited inputs from the spreading grounds, so modeling other runoff-capture infrastructures was a largely theoretical exercise. When engineers modifying his tool for the SCMP wanted to model smaller-scale infrastructures they might build in the future, James advised them to add lots of "little spreading grounds" to their simulations and see what happened. But he was blunt about the fact that an extensive new program of monitoring would be necessary to meaningfully calibrate the models and reflect the volumes of water that installations absorbed in practice.

Establishing such a program, however, is easier said than done, due in part to the dearth of widely accepted infiltration monitoring techniques at the distributed scale. In early 2015 Nikola, a staffer at a local environmental NGO, wrestled for months with the difficulties of developing such a protocol for domestic rain gardens. Funded by a grant from a water wholesaling agency, his organization was installing and monitoring mulch-covered stormwater-capture swales on residential properties, aiming to quantify both the reduction in potable water use (due to not needing to water a lawn) and the infrastructures' recharge capacity. The group relied heavily on volunteer labor to keep installation costs down, so I spent a sunny Sunday morning working alongside a rotating crew of neighbors and university students to help plant the newly terraformed front yard of one of their sites. To ease the process in the hard soil, a couple of burly interns were operating mechanical hole-digging machines. This meant that much of the volunteers' time was spent sheet-mulching areas that had recently been covered in grass to prevent its regrowth.

During a break (held on the backyard's intact lawn), I asked Nikola a few questions about monitoring as we stood in line for some homeowner-provided lunch, and he suggested a follow-up call to get into the weeds of it all. When we reconnected, I learned that at that house, the data gathering would rely on a sub-metered sump pump installed in a five-gallon bucket at the deepest part of the rain garden. This approach hadn't been the NGO's first idea; for weeks they had been set on creating a tiny weir system at the edge of the depression, but during implementation it became clear that the approach would be unwieldy and impractical. Nearing the project's end, he admitted that for some properties, they had abandoned monitoring efforts altogether: "We had to go back to older methods, taking areas to calculate, using runoff coefficients to model." Such stories of frustration seemed to signal that for all its acknowledged data gaps and deficiencies, infiltration modeling would remain a key mode of representing infrastructural nature's water-supply contributions for the foreseeable future.

Nick and Ellen's debates with the LADWP engineers reveal a dimension of this process that rarely surfaces in its official documentation. Namely, these characterizations of local stormwater flows are built on a broader range of inputs than is commonly acknowledged, and even minor tweaks to any one of the numerous variables can lead to meaningful changes in the estimates these models produce. Armed with detailed site plans of their installations and months of firsthand observations, the NGO workers could recount the trajectories of specific rainstorms that contradicted the assumptions underpinning their water agency colleagues' model. They frequently invoked the downpour that had toppled Woodman—categorized as a "five-year storm," an event on a scale that forecasters assumed to happen at least once every five years—to discuss the effectiveness of the oversized rain garden they had dug in front of one house. During the deluge, a rerouted downspout directed the entirety of the roof's runoff to the swale, which never overflowed. Based on this experience, Ellen and Nick termed it the "five-year rain garden" and pushed hard to ensure that the water agency's calculations reflected this impressive capacity for water capture.

As such negotiations attest, the hydrological models served as key mediating sites for stakeholders, terrain where competing accounts of nonhuman nature and green infrastructure were rendered (at least somewhat) commensurate. Once absorption figures were successfully brokered in such settings, the numbers characterizing an installation's runoff capture capacity could circulate in water agency presentations and NGO promotional materials, lending weight to claims of infrastructural efficacy. The stakes of this work become clearer in debates over how, exactly, value should be ascribed to green infrastructure installations and the stormwater that they infiltrate. Metrology is essential to and entangled in the process of producing ecosystem services as investable quantities, as they must be "known, counted, expressed in standardized units, and, ultimately, made commensurable with monetary value."[19] As the next section demonstrates, state-led processes of such commensuration also offer opportunities for strategic advocacy on behalf of infrastructural nature.

VALUING THE RAIN—AND THE OTHER BENEFITS OF THE SWALES THAT CAPTURE IT

"What do we value and what does the future of successful stormwater capture look like?" Our facilitator, reading from her PowerPoint slide, addressed this question to the roughly two dozen attendees of the morning's Stormwater Capture Concepts Charette. Cosponsored by the Bureau of Reclamation and the LA County Flood Control District, the gathering was staged to solicit input from key stakeholders to help the agencies advance a planning study on stormwater in the LA Basin. Hosted in a chilly, cavernous room on the ground floor of Metropolitan's headquarters, the meeting had drawn a mix of NGO and water agency workers. As latecomers trickled in from nearby Union Station, the assembled group listened to a brief presentation about the convenors' climate change modeling work, which predicted even more extreme local hydrological variability in the years to come. Now, though, it was our turn to talk, and our hosts' opportunity to document our comments for future reference in their official report.

Ellen broke the silence. "Value, I would say go 180 degrees away from the current value of stormwater. Our systems and city are built to fear it and throw it away. . . . I would reverse that and retrofit the city—because cities are always being rebuilt—and value it and recognize the relationship between land and water." Eric, a middle-aged white man who worked at a different NGO and was another fixture at such events, went next. Building on Ellen's remarks, he urged the group to think with an expansive notion of value. "We should value it as an opportunity and make sure that the success we achieve is one that brings as many benefits as possible, that we build a system that doesn't just do flood control, but that brings parks and jobs and resilience. We talked about wanting to build resiliency, but most important is building a multi-benefit system." Hanna, a younger Latina woman and a longtime staffer at a third NGO, followed up, articulating what I understood by then to be the trio's animating aim. "We should value the diversity of capture methods, be sure that we value distributed capture as much or more than we do centralized capture. . . . I think the success that would come out of that would be a greater reliance on local supply."

Considered together, these comments suggest the contours of the three advocates' shared project. As detailed earlier, they are working to legitimate an account of distributed green infrastructure's efficacy as a runoff capture technology. But as Ellen's and Eric's words signal, they are also making claims about existing and potential connections between stormwater and several notions of value, none of them stably linked to price at the time of that meeting.[20] Characterizing runoff in such terms, the trio drew attention to the limitations of established practices of pricing the work of land-based stormwater capture, strongly suggesting a need to rethink it.

Within the geographical literature, such attempts to assign value to ecosystem services are typically glossed as defining elements of the commodification of nonhuman nature.[21] However, most scholars are also quick to acknowledge the fact that such services are in many ways an awkward fit for the commodity form, winning them the labels of fictitious (in the Polanyian sense) or "frictitious" commodities.[22] As geographer Karen Bakker demonstrated in her foundational accounts of water as an "uncooperative" commodity, the complex, inconstant materiality of the substances, flows, and ecologies involved is a key source of such categorical tensions.[23] It is true that in some circumstances (in particular, bottling), water can be made to look and circulate like the wood-turned-table that Karl Marx used to sketch his influential account of the commodity form. But within urban water networks, the substance has a differently situated relationship with price—one conditioned by the costs of the infrastructure that moves the liquid and the resource's availability within the local environment. Given all of this, it is perhaps unsurprising that stormwater, a widely dispersed, unpredictable, often ephemeral form of water, presents such difficulties for both calculable capture and valuation. Further, the sort of nature-based installations proposed for its infiltration in the northeastern Valley do not fit comfortably within "grey" paradigms for infrastructural asset valuation and management, presenting additional accounting complications.[24]

In the LA context, the SCMP was meant to help resolve such ambiguities. The consultants and agency staff who worked on the plan were tasked with developing analysis that established price-per-acre-foot estimates for newly captured stormwater and a cost cut-off for priority projects. The modelers derived those figures from a combination of factors: new infrastructure construction costs, anticipated maintenance expenses, localized rainfall predictions, and analyses of soil and aquifer characteristics. The final product was a set of price points, numbers that the plan's authors hoped would circulate without much difficulty. Several consultants and LADWP staffers involved in developing the SCMP told me that they believed the document's stormwater accounting methodology, rather than its long list of proposed projects, would be its most substantive contribution to the future of runoff infiltration within the city.[25]

Many of my interlocutors, however, were unimpressed by these outputs. Notably, this group included skeptics who felt that the SCMP's figures projected the cost of stormwater capture to be artificially low. A primary concern among this group was that the authors' reliance on annualized averages to predict rainfall within the city failed to reflect the unreliability of this prospective resource. As they noted, in addition to the possibility of intense storms overwhelming infiltration infrastructures (especially smaller ones), yearslong droughts were also always a possibility within

LA. During an especially dry year, infrastructures assumed to infiltrate ten acre-feet of runoff could plausibly absorb less than one.

Recognizing these conditions, it becomes clear that by investing in the Woodman project, for instance, the LADWP helped to finance an array of berms, swales, plants, and curb cuts, not any kind of guarantee that a particular volume of water will soak into the San Fernando Groundwater Basin (or for that matter, that the infrastructure will remain intact). The price of infiltrated Woodman water produced by the SCMP modelers is thus an informed guess about the future of rainfall and water flows within the northeastern Valley. That figure is tagged to the only certain number for the agency's ledger—the money it poured into the installation's design and construction—circumstances that underline the uncertain relationship between investing in and realizing new water supply via recharge. As one city engineer described his hesitations: "Recycled water is much more reliable than stormwater, or cisterns or MET water, for that matter. I don't know what premium we should put on reliability, it's unknown but could be substantial."[26]

While acknowledging such concerns, green infrastructure advocates argued that the bigger problem with the SCMP was its failure to properly account for the full suite of benefits green infrastructure installations provided. Eric's line at the charrette about the potential for better local stormwater management to bring "parks and jobs and resilience" hints at the range of services advocates understand as potentially connected to these installations. Grounded in the recognition that these nature-based infrastructures transform slices of the urban landscape, this framework for valuation demands a more holistic evaluation of those slices of land—which, proponents maintained, would ultimately reveal them to be quite a good deal for the city. Converting an abandoned parking lot into a small park terraformed with stormwater swales and berms, for instance, presented the opportunity to do more than enhance water supply; it could provide green space for residents, reduce (at least slightly) flood risk during downpours, and offer employment to construction and eventually landscaping workers. While proponents were more explicitly attentive to material flows than social formations, their emphasis on such infrastructural nature's capacity to support benefits like "parks and jobs and resilience" signals a distinctly relational approach to the pursuit of the desired ecosystem services.[27]

As Ellen's, Eric's, and Hanna's comments at the charrette suggest, stakeholder input sessions held as part of agency planning processes served as forums for advocates to formally register these assessments of infrastructural nature's "true" economic value to the city. At a follow-up meeting for Reclamation and Flood Control's ongoing study, I watched NGO and city staffers alike stew in frustration as an engineer from the project presented his team's preliminary assessments of different stormwater capture projects' desirability and value. When he paused for questions, a staffer from Sanitation launched into an extended critique, arguing that they had severely undervalued the multiple benefits small-scale green infrastructure installations provide. The project team's next speaker, an economist, did not fare much better. The month before, he had circulated to key stakeholders a proposed methodology for assessing the benefits different stormwater infrastructures provided: a sheet listing many categories of "benefits" that instructed them to rank each benefit on a scale of one to ten for a project with which they were familiar. The project team would be conducting a similar rating exercise of their own but would plan to take the opinions of all respondents into account.

Sitting in the NGO's office when Ellen opened the email, I got an earful about the deficiencies of this approach to benefit characterization. Judging by the stream of withering comments that filled the stakeholder meeting room after the economist's presentation, the others clearly shared her assessment. Several suggested that the categories were so broad as to be essentially meaningless. Others expressed frustration at the failure to explicitly include ecosystem functions or to address the greenhouse gas emissions embedded in concrete infrastructures. The economist forged ahead good-naturedly, acknowledging the limitations of his methodology but also attempting to situate it within the state of economic and policy literature on how to assign monetary value to benefits like increased water supply, ecosystem services, and recreational benefits. His conclusion was that there was not yet a broadly accepted approach to comprehensively pricing the benefits associated with stormwater capture (nature based or otherwise)—hence, his work entailed substantial invention under major time and resource constraints.

Observing the action, I sympathized with this challenge. Months earlier, Ellen had tasked me with conducting a quick literature search on current ecosystem valuation practices to inform a public comment she wanted to submit on an earlier draft document from the same study. Like the economist, most of what I turned up seemed piecemeal, partial, and unstandardized, often relying on small-sample surveys of respondents' "willingness to pay" to derive valuation figures for protected and restored landscapes.[28] Like the measurements and models of ecosystem function discussed previously, the practices of pricing the same phenomena were far less comprehensive than some accounts of the creeping commodification of nature in the critical social science literature suggest.

For my interlocutors, these gaps presented an opportunity to nudge the methods underpinning agencies' stormwater analysis toward, a more capacious accounting and valuation of nature-based infrastructures. Long after the official end time of the meeting discussed earlier, Ellen stayed on to continue reviewing potential approaches with the project team. As in the modeling conversations, she carefully questioned the assumptions underlying the chosen methodology, attempting to reframe the boundaries of the system they sought to characterize.[29] Such protracted negotiations underline the contested, capacious notions of value entangled within the project of replumbing the city. But such work, while infrastructural to a long-term program of developing infrastructural nature, could do little to address questions of jurisdiction and fundability that frequently stymied efforts to get new projects in the ground. As the next section shows, my interlocutors drew on a different repertoire of narrating nature-based infrastructures in the service of creating the necessary links.

LINKING FUNDING WITH INFRASTRUCTURE

Over the course of my months in the field, I watched Ellen guide many visitors through her NGO's infrastructural nature pilot project in the northeastern Valley. The composition of our tour groups varied: curious college students; staffers from other local NGOs; workers from city, county, and state agencies; and visiting engineers from across the region all walked the neighborhood with her during my time conducting participant observation within the organization. Regardless of who was attending, these were usually pleasant excursions, buoyed by Ellen's enthusiasm and carefully crafted to convey her vision of the wide range of benefits that reworking one's yard to maximize stormwater capture could bring to a neighborhood.

Typically, the tours would begin with the group gathering in front of a particularly charming house on a block with a handful of other retrofitted properties. Several of the yards on the street featured rain gardens, planted swales constructed to absorb the flows from a roof's gutters, and parkway basins, depressions dug into the grassy strip between the sidewalk and the street designed to siphon in runoff from the road. Ellen would explain the water infiltration benefits from these installations in expansive detail, always with reference to the aquifer beneath our feet. But she also made sure to draw our attention to other aspects of the retrofits. Selecting flora to plant in and around the swales, the project team had chosen as many native plants as possible. Bright orange California poppies caught everyone's eye in the spring, while the fragrance from white sage bushes often drew appreciative comments. If no one else mentioned them, Ellen would be sure to note the concentration of bees and hummingbirds circling these plants. Narrating these details, she was making certain that her audience appreciated the range of local environmental amenities and entanglements that such home-scale retrofits brought to the neighborhood. Through the domestic landscapes of the pilot project, she was telling a story about the range of ecological services that an extensive program of distributed stormwater capture could bring to the terrain of the northeastern Valley—if only public agencies would collaborate to invest in it.[30]

Such tours are particularly useful for demonstrating the more-than-human collaborations involved in advocating for distributed infiltration infrastructures. Like critical social scientists attuned to how direct engagements with the materiality of domestic gardens can produce social and political-economic relations, Ellen understood visits to the neighborhood as opportunities to solidify new notions of local material flows, a chance for the sights and smells of the reworked yards to help transform visitors' understandings of urban land's potential to support a range of ecosystem functions.[31] Presenting such quotidian, aesthetically appealing slices of metropolitan land as functional water supply infrastructure allowed her to begin complex discussions about the appropriate scale and configuration of twenty-first-century urban water networks.[32] Seeing, smelling, and hearing the yards' berms, swales, sage, poppies, and bees was central to this process; the sensory appeal of the gardens did a lot of the work for her, Ellen often told me. But the careful framing her narration provided was also key to directing visitors' attention and responses to the landscapes.

Listening along as Ellen guided a pair of staffers from a state agency through the circuit one day, I noticed that she appeared to be especially on her game—which seemed like a good thing, given the potential stakes of that walkthrough. The two women had flown from Sacramento to LA for a brief listening tour among public agencies and nonprofits working on local stormwater issues. In the months ahead, they would help craft the language and rubrics to guide the disbursement of $200 million in storm-water project funding from California's 2014 water bond. Before doing so, they wanted to talk with practitioners and see how already-built projects were working. Ellen, who hoped to replicate the retrofit project in more neighborhoods, understood that conveying her account of what sort of stormwater capture installations are valuable (and why) to this audience could be vital to realizing that goal. Standing in front of the house with the "five-year rain garden," she recounted a wet day the previous fall when a staff member took a cell phone video as he walked down the sidewalk. Sheets of water were sliding off the lawns of the neighboring (nonretrofit-ted) homes, she explained, but when he got to this house, the sidewalk was just damp. All the runoff on the property had been absorbed by the swale. "The difference was amazing!" she exclaimed, then pivoted to a discussion of how these yards now function as excellent pollinator habitat.

Over burritos back in the office, the guests listened attentively as Ellen listed ways that the language in their grant solicitation could be structured to support future projects like the one that they had just viewed. They thanked her for the suggestions and complimented her passion as they left for their next meeting. In a quick debrief, Ellen, another staffer named Nancy, and I agreed: they seemed to get it, to grasp the recharge capac-ity and all the associated cobenefits that Ellen saw in the installations. But, we acknowledged, we had no idea if this understanding would ever translate into funding for future projects. Attracting the necessary capital to develop distributed stormwater infrastructure across the northeastern Valley, I had come to appreciate, required sustained negotiations over the nature, contributions, and value of those installations with a wide range of people—long-term efforts with no guarantee of success.

These difficulties were structured by the divergent remits of the agencies involved in managing the flow of water to and through LA. Over the course of the twentieth century, water supply, sewage management, and flood control emerged as distinct governance challenges, thought to be best addressed by dedicated agencies with distinct mandates. As others have demonstrated, this institutional legacy now complicates the city's efforts to manage stormwater as a resource rather than as a hazard.[33] Grounded in the grey epistemologies that privilege centralized water infrastructures designed to "isolate water and manage it largely separate from social systems and the messiness of day-to-day human interactions," this bureaucratic configuration is not well equipped to finance distributed or nature-based infrastructures.[34] The difficulties of agency funding coordination are widely acknowledged within the city's bureaucracy and NGO community and cited as the organizing rationale behind LA's One Water 2040 planning process, which attempted to bring together representatives from across agencies to establish a comprehensive approach to the city's flows. Published in 2018, the report—more than four thousand pages in length—sketches detailed plans for rationalizing all the water within LA. Like many such aspirational planning documents, however, the One Water plan does not fully resolve the differing agency remits that so stymied the work of drawing funds to distributed stormwater projects during my research period.

Such gaps were readily apparent in discussions of infrastructural nature's multiple benefits. Keith, a retired white businessman and vocal participant at both RWAG and One Water stakeholder meetings, was particularly pugnacious in policing the scope of LADWP plans and projects. "Every time they talk about mutual benefits, I hold onto my wallet," he told me during a lengthy Saturday afternoon chat in his mansion's well-appointed living room. As a ratepayer, he hated the idea of the water utility putting money from his monthly bill toward anything other than sustaining the city's water supply. It wasn't that he was *against* realizing cobenefits associated with green infrastructure installations, such as improved water quality in the LA River and its tributaries (a goal mandated by the federal Clean Water Act and a frequent cojustification for establishing green infrastructure programs). He just thought it was inappropriate for a water utility to ask consumers like him to underwrite those outcomes through increased water rates. "Sanitation has the primary responsibility for water quality," he accurately noted. As such, he wanted the LADWP's cost share of projects providing water quality benefits pegged tightly to the water supply benefits. "Should it be a 50/50 split?" he asked rhetorically. Probably not, because, he concluded, the LADWP is "not the deep pocket [to pick]."

In meetings and interviews, city staff often expressed more flexible positions than Keith's on such points. "It's easy to use standard economic valuation methods," I once heard an LADWP engineer muse to a stakeholder group, in response to an outburst from Keith demanding a greater reliance on such forms of accounting in making stormwater funding decisions. "But it can be short-sighted not take into account the way the future could be dramatically different, particularly when it comes to drought." In practice, however, the agency appeared determined to limit its monetary contributions for new stormwater infrastructures strictly to those deemed cost effective from a water supply perspective, based on its modeling efforts.[35] As such, funding arrangements like Woodman's—cobbling together monies from the LADWP, a state grant, and Sanitation—were frequently necessary to realize such distributed recharge projects.

These circumstances meant that NGO workers and city staff alike were constantly searching for new pots of money that might provide the matching funds necessary to advancing green infrastructure design and construction work. Rick, a Sanitation staffer, and I discussed this pattern while shuffling slowly down a two-block stretch of unmemorable strip malls one mild spring afternoon. That bit of road had recently been designated one of several potential recipients of revitalization money from a private funder within the city. Learning about the funds, a group of colleagues working on stormwater issues from a mix of organizations had decided to do a walkthrough of the area, to see if it might have any sites with runoff capture potential. The best-case scenario: we would find some, a couple of participants would convince their organizations to collaborate on developing a grant proposal to underwrite part of the design and construction of green infrastructure installations, the application would win some money from the funder, and all relevant organizations would then throw in the necessary contributions and the facilities would get erected.

Outlining this course of conditional events, Rick couldn't help but laugh. He estimated with a shrug that in his years working for the city, about one in four projects he had helped design was eventually funded and built. And the ratio was no better for stormwater infrastructures, he assured me. Trained as a landscape architect, he was conversant with the paradigm of nature-based solutions and appreciative of their multiple benefits. From his vantage, LA's public agencies just weren't set up to realize them. Many shared this perspective but persisted even so, seeking out ever more creative sources and combinations of project funding, writing grant after grant, and—as in Ellen's case—conducting dozens of pilot project tours, attempting to mobilize already constructed infrastructures in the pursuit of money for additional installations.[36]

Such ongoing hustles for funding are, of course, unexceptional in many NGO and public agency settings. But they are noteworthy in the context of LA's stormwater infrastructure trajectory because they illustrate institutional factors that can quietly confound the translational efforts this chapter outlines. Indeed, as geographer Joshua Cousins puts it in his writing about the LADWP's stormwater modeling: "These calculations, measurements, and inscription devices have material effects, which influence the ways stormwater circulates as a metabolism through Los Angeles" (2017c, 376). These effects, however, are situated within structural arrangements that can delimit or direct that influence. As such, the work of narrating for a diverse audience of potential funders the ecosystem services infrastructural nature can provide emerges as labor that underpins the rerouting of LA's stormwater metabolism. While neither predictable nor always successful, my interlocutors' efforts to mobilize models, valuations, and embodied experiences of these infrastructures clearly play a role in reshaping the city's runoff flows.

At a community forum held in a hospital complex close to the Woodman Median, I watched the crowd as Ellen narrated the local hydrology. "Show of hands: who lives in the Valley?" Several dozen arms shot up. "Great. So you are our water future. We have the best possible water storage for LA here, beneath our feet, we don't need to build it." She proceeded quickly, flashing images of the home-scale green infrastructure installations that her NGO had recently completed on the screen. "With these, you can help get the water from your roof—you know, the stuff that normally ends up in the gutter—into that basin."

Clocking the confused looks on a few faces, she elaborated, situating the planted swales pictured on the screen in the context of the northeastern Valley's porous soils and the LADWP's groundwater wells. The befuddled expressions faded. By the time the talk gave way to a brief Q&A, several attendees were intrigued enough that they asked how they could support her organization's work. Infiltrate water on your own property—and write to the LADWP to get them to fund projects like this one, she replied. Nods all around. Though the likelihood of any individual following through was low, the work of broadening the constituency for distributed stormwater infrastructure appeared largely successful that day. Given the abstract nature of any given front-yard rain garden's contribution to the municipal water supply, Ellen's artful framing of the material linkages involved was clearly central to the attendees' reassessment of the ecological potential embedded in the soil beneath their feet.

As this chapter has shown, such discursive infrastructural labor is central to my interlocutors' efforts to reroute LA's stormwater flows, a quietly constitutive form of adaptation work. Because the distributed infiltration installations that they seek to develop fit uncomfortably within established paradigms of knowing, valuing, and funding urban nature and infrastructure projects, advocates undertake long-term efforts to make such facilities legible and desirable to key audiences. State-led practices of environmental modeling and accounting are key terrain in these processes, serving as sites of negotiation over the nature and value of stormwater and green infrastructure alike. But while much of this work is directed toward the relatively small community of experts and bureaucrats typically tasked with managing the city's waterscape, Ellen's presence at the northeastern Valley community forum suggests the perceived importance of a different subset of Angelenos: those living on the porous soils that overlay the San Fernando Basin. The next chapter explores the role scripted for those communities in the infrastructural arrangements the advocates seek to produce, highlighting another, far more dispersed form of labor that this approach to replumbing the city would entail.

5 Ecosystem Duties and Environmental Justice

Kneeling alongside Nancy, Salvador, and Juan while struggling to unroot a seemingly immovable weed, I was once again reminded that, even long after funding is procured and construction is completed, green infrastructure demands a lot of human labor. The rain garden that our quartet was tending that cloudy Friday morning in the northeastern Valley had been carefully designed to absorb the runoff flowing from the roof of the modest bungalow sitting a few yards away from us. Built as part of the neighborhood retrofit project, the installation had descended into disarray in recent months, filling with weeds and leaves and detritus that limited the volume of water that the swale could hold. Like a water supply reservoir accumulating silt, the shallow depression's unruly plant growth was compromising its intended infrastructural functionality.

The homeowner's hands-off approach to maintenance had spurred anxious discussion within the NGO in the weeks leading up to that morning, eliciting winces and worried commentary when staffers passed through the neighborhood. Outreach to the household produced little change. Eventually the yard's deterioration resulted in the decision to send Nancy, a long-time staff member, over to the house to lend a hand. An affable, chatty, middle-aged white woman, Nancy loved gardening, enjoyed spending time with the pilot project participants, and seemed happy enough to pass a day in the mulch. Aware that local schools were on spring break that week, she had cajoled Salvador and Juan, Latino teenage brothers from another participating household, into joining her. With my schedule that day open until an afternoon interview, I offered to pitch in for a few hours. So we dug together, joking occasionally about the romantic lives of bugs as we filled a growing mound of garbage bags with the weeds and bits of trash.

The work wasn't easy, but the camaraderie was pleasant. The boys, thanks to their hours participating in the NGO's educational workshops and helping to maintain their own home's retrofitted yard, knew far more about the rain garden's plants than I did. They liked California poppies, hated earwigs, and gently redirected my efforts when I started tugging on something that was meant to stay in the ground. When Juan's eyes started watering from an allergic reaction caused by one of the plants he had been handling, he jogged home to wash his face and returned with an armful of water bottles sweating from the fridge. Not long after, a shouted "hey!" interrupted our sweaty reverie. A couple of men in a Bureau of Sanitation truck waved and offered to haul away our refuse. We gladly accepted and watched as they tossed the bags into the waiting truck, acutely aware of the effort they were saving us.

On many registers, these muggy, mucky hours were unexceptional ones within LA's urban fabric. This is a city dominated by single-family homes, most with yards that require some form of ongoing upkeep. Whether undertaken by household members or landscaping contractors, this work is widely considered essential to sustaining the aesthetic norms and property values of Southern California homes. But unlike most LA yards, the one we were maintaining that day is understood—thanks to its enrollment in the city's efforts to recharge the San Fernando Groundwater Basin—as not just a visibly prominent element of someone's private property, but also water-providing infrastructure for the public grid. And as such, the work of its upkeep raises questions related not only to aesthetics and real estate–based wealth, but also to obligation and fairness and human labor in the context of initiatives seeking to cultivate the work of urban nature in the name of climate adaptation. The uneven distribution, naturalizing representations, and tactical nonvaluation of such hybrid labor are this chapter's core concerns.

As with the domestic reuse systems considered in chapter 2, examining the localized environmental practices associated with distributed recharge installations reveals understandings of connection and civic duty that proponents of LA's infrastructural nature projects seek to incite among residents. The northeastern Valley's green infrastructures are designed to maximize runoff infiltration, materially linking everyday spaces like street medians and front-yard gardens to the aquifer that sits beneath them. In doing so, these installations enable such small, multipurpose slices of urban land to contribute volumetrically to LA's collective resource base. Maintaining the infrastructures' functionality thus emerges as labor that helps to augment not the individual's domestic supply but the entire city's water stockpile.[1] Like the work of managing LA's aqueducts, reservoirs, and other more traditional water provision infrastructures, these are exertions with ramifications that stretch far beyond the installations themselves. But unlike maintaining the city's established network of pipes, much of this labor is assumed to be freely or cheaply given, primarily by residents of the northeastern Valley. Such arrangements of unremunerated work mark many projects of green infrastructure development, ecosystem service provision, and even general programs of urban greening across the globe, signaling their stakes for contemporary efforts to manage metropolitan environments.

Ecosystem duties, the term I develop to characterize the human exertions necessary to maintain these assemblages, intentionally evokes the more-theorized category of ecosystem services to highlight the work of people that the realization of such ecological functions can entail. Tracking when and how this labor becomes enrolled in LA's network of water management, I show how the exertions tactically excluded from valuation are in fact constitutive of these systems of resource provision and the broader political economic arrangements that they underpin. Further, I consider how, within LA, these new forms of work spatially reinscribe the city's classed and racialized patterns of socio-environmental inequality. Recognizing these dynamics suggests the value of incorporating analyses of such forms of ecological labor into considerations of environmental justice.[2] This move to foreground peoples' contribution to these configurations of hybrid labor should not be read as a rejection of the scholarship that decenters human actions and intentions in the production of socio-ecological arrangements, or an endorsement of the notion that other-than-human nature (infrastructural or otherwise) is inert and infinitely malleable. Rather, I seek to improve the precision with which such collaborations are conceptualized, highlighting the stakes of what, exactly, is expected of whom in projects that explicitly script such active roles for other-than-human nature.

As I show, notions of relationality and obligation (in addition to economic necessity) are central to the promotion of these arrangements. The work of infrastructural nature maintenance is frequently framed in terms of residents' "stewardship" of LA's land and water. Sustaining green infrastructure thus emerges in discourse around such installations as an ethically and affectively laden practice, exertions with the potential to contribute to collective flourishing, deepen one's connection with the local environment, and assuage diffuse ecological anxieties. Such framings resonate with Scaramelli's notion of a moral ecology, as did Issa's thoughts (quoted in chapter 2) about her greywater systems and practices.[3] This concept conveys the sense among many of my interlocutors that nature-based solutions not only were technically efficacious but also held the potential to spur more pleasurable modes of living in the city while addressing long-term environmental inequities within the urban landscape.

Narrated as installations that could bring a range of ecosystem services to marginalized neighborhoods, proponents thus presented green infrastructures (and the work of sustaining them) as assemblages that helped to ameliorate decades of localized environmental degradation and neglect, in addition to augmenting municipal water supply. In practice I observed these entanglements to be highly generative, but on a more ambivalent register than their proponents typically suggested. Senses of pleasure, connection, and pride in one's role in the city's environmental fabric were often part of the story. And so were expressions of frustration and resentment at the burden of managing the (often unruly) infrastructural nature. Such accounts of unpaid ecological labor signal the frictions that attempting to incorporate it into the city's waterscape can entail and its complex articulation with understandings of environmental justice.

The analysis that follows foregrounds experiences in and around the neighborhood pilot project referenced earlier (and throughout the preceding chapters). The project sought to demonstrate the effectiveness of

intimately scaled water-capture infrastructures, in the service of expanding their uptake to address a range of water supply and quality challenges
facing the city. While the participants, most of whom both owned their
homes and identified as nonwhite, did not have to pay for the retrofits
(and generally described these free home "upgrades" as a motivation to
join the project), they did sign paperwork committing to maintain the
installations on their property for three years (notably, some similar city
projects require a twenty-year commitment). This work entailed regular
clearing, desilting, weeding, and pruning of the infrastructures, which, as
I have mentioned, included a mix of parkway swales, rain gardens, rain
tanks, and permeable driveways. As chapter 3 discussed, the targeted
neighborhood lies within a part of the city that public agencies have long
treated as a low-priority landscape and allowed to flood regularly and accrue disproportionate pollution burdens. The recent influx of public sector attention to the area—driven by water manager and environmental
NGO assessments of the area's hydrological characteristics—thus marks a
departure from past patterns of state neglect. I show in this chapter how
the realities of the maintenance work that proponents of these infrastructural natures seek to mobilize further complicates this shift in attention.

ECOLOGICAL WORK OUTSIDE OF VALUE

Elaborating the notion of ecosystem duties through an analysis of these
distributed green infrastructure projects extends anthropological con-
ceptualizations of the relationship between infrastructural natures and
human labor. Other ethnographic case studies of infrastructural nature
have focused on rural PES schemes that script a substantive role for hinterland residents in the realization of the desired ecological functions.[4]
Other-than-human nature, in such arrangements, is understood as only
capable of providing its valorized services through the construction, planting, culling, or guarding efforts of rural communities.[5] The scale and duration of payment for these efforts can emerge as a key site of negotiation
between funders and workers, highlighting the projects' contested role
in sustaining rural livelihoods and facilitating new claims on the state.[6]
In contrast, LA's approach to distributed infrastructural nature proceeds

from the assumption that much of the necessary human work can be mobilized for free. This is a widespread paradigm for siting and sustaining
urban green infrastructure both within and beyond the United States.[7] As
such, the weedy swales of the northeastern Valley present an opportunity
to consider how and why ecological labor so frequently emerges as work
with no link to livelihoods or remuneration within such arrangements, as
well as the stakes of such configurations. Situating them within the larger
arena of work (human and otherwise) outside of capitalist value clarifies
some of the key contours.

   Though critiques of the ecosystem services paradigm have raised concerns about the creeping commodification of some forms of nature's
work, it is increasingly clear that in practice, infrastructural nature projects often rely heavily on the nonvaluation of many forms of labor.[8] In
Marxian terms, we might put it this way: while exchange value is typically
produced through labor, not all labor produces exchange value. Feminist
theorists have long drawn attention to this point, highlighting the unwaged domestic and reproductive labor that underpins capitalist political
economy.[9] Building on these insights, geographer Jason Moore engages
Marx to build a robust argument for considering the selective omissions of
this and other subsets of human labor on the same plane as the "work" of
nature.[10] Recognizing such constitutive exclusions from the process of valuation suggests the importance of attending to the ways that these forms
of human work get ideologically coded as "natural" and of considering
their relationship to the un- or undervalued work of nonhuman nature
under conditions of capitalism.

   The anthropological literature on the labor of social reproduction foregrounds the ways in which gender, race, and class shape exclusions of
human labor from economic valuation, within and beyond the household.
Classic texts explore how the work of child-rearing and other domestic
labor can come to be understood as "natural," feminized acts of care and
code the home as a realm beyond the market, a pattern that has been
shown to extend to home-scale environmental provisioning work.[11] Recent ethnography within the LA context highlights the stickiness of these
categories, elaborating how hiring paid care work creates friction among
both wealthy white women and the less-affluent women of color that they
frequently employ for this labor.[12] Though attentive to the possibilities

for change and "revaluation" within these configurations, particularly the ways in which neoliberalism has refigured them, this body of work tends to foreground how capitalism was built (and continues to rely) on the free, feminized work of social reproduction.[13]

These insights articulate with research from Black studies on the constitutive nature of unfree and unwaged Black and Brown labor in the production of the modern global economy.[14] This literature also details how even the most grueling forms of labor carried out by Black and Brown workers have been naturalized in a range of exploitative contexts, from plantations to prisons and beyond.[15] Taken together, these threads of scholarship highlight the need to consider carefully *which* people might be pulled into the work of helping provide ecosystem services (particularly in the residential context) and how such arrangements map onto existing patterns of difference and inequality.

Engaging these themes, writings within environmental anthropology highlight how, labeled labor or not, more-than-human collaborations in pursuit of landscape's health or restoration rarely result in straightforward exchanges of wages for peoples' exertion. Characterizing these efforts as "feminized, racialized, and naturalized," much like other forms of care work, Danielle DiNovelli-Lang and Karen Hébert use a case study of an Alaskan forest-repair project to show how such programs exploit the labor of communities marginalized by the very extractive industries that created the need for such remediation.[16] There are parallels between these communities and the neighborhoods most targeted for green infrastructure pilot projects in LA, where residents who have long suffered the effects of the state's underinvestment in their local environment are now being drawn into "stewarding" water-capture installations. Other cases in the literature emphasize the affective pleasures scripted for the humans involved in such undertakings, flagging the tendency to present them as noble, fulfilling pursuits for paying volunteers.[17] The reversal inherent in such projects—paying to exert, rather than exerting to be paid—underlines the scrambled relationship between human effort and economic value that marks many of these arrangements.

These insights highlight the growing salience of labor for understandings of environmental justice. For decades, environmental justice scholars and activists have demonstrated how racial capitalism produces uneven exposures to environmental harms and risks, such as toxic waste dumps, lead-tainted water, and polluting industrial facilities.[18] Complementary analyses foreground how communities marked by racial and economic privilege frequently enjoy disproportionate access to environmental amenities, such as clean air and water and green space.[19] Recent, related work on so-called green gentrification has shown how projects of urban greening can become imbricated within processes of dispossession, raising property values and forcing out longtime low-income residents.[20] My project here is to extend the frameworks this body of research has developed by incorporating the long-term, embodied work of environmental production and reproduction into analysis of such well-intentioned projects of urban ecological improvement. Approaching infrastructural nature installations as sites of hybrid labor clarifies their multivalent role within the urban fabric.

MAINTENANCE LABOR, DEVOLUTION, AND TACTICAL NONVALUATION

Virtually all infrastructural networks require sustained human work to remain functional.[21] Often unevenly distributed, favoring privileged communities over marginalized ones, state programs of infrastructural maintenance can serve as potent sites of politics, techno-political terrain where communities struggle to access the promises of citizenship.[22] Bearing in mind that beyond improving conditions for residents, fulfilling maintenance commitments can also serve a key legitimizing function for state agencies, it is notable that the distributed infrastructural nature projects examined here are explicitly premised on public agencies playing a minimal role in upkeep.[23]

I got an even clearer sense of the considerable upkeep green infrastructure requires from pilot project participants Ned and Nora, a fortysomething white couple who took great pride in their up-to-date efforts. They walked me through their property during our interview, detailing the retrofits and describing a regular regimen of pruning and weeding, watering plants during the long dry season, scooping trash and mucky buildup from swales after each rainstorm, degunking the permeable driveway, and

Figure 16. Carefully maintained green infrastructure installation. Photo by author.

Figure 17. Overgrown infrastructure installation clogged with refuse. Photo by author.

frequently clearing gutters to ensure that all rain flowed from the roof into the intended basins. When I asked, they estimated that this all took several hours a week, more during the wet season when rains and runoff dragged in debris that clogged the trenches. They acknowledged that it was more work than using automated sprinklers and mowing the patch of grass they had had before, but both felt compelled to keep it up to ensure that the installations continued functioning as intended. The project team, they told me, had emphasized that overgrown, trash-filled green infrastructure will not absorb nearly as much water as carefully maintained installations. Though the couple did not use the language of "duties," their sense of obligation—to the local environment, the city's water supply, and the project—was evident throughout our conversation.

NGO and city employees who worked to spread such nature-based recharge installations are aware of the labor that their maintenance requires. Notably, while willing to advocate for the economic value of the work green infrastructure provides—in terms of floods prevented, streams cleaned, and groundwater basins augmented—in many contexts these proponents sought to minimize the same form of value when it comes to the maintenance humans do. "My vision of maintenance is all about money," Latif, a manager at LA's Bureau of Sanitation, told me over coffees one morning. "My vision of maintenance, at the end of the day, is that it should be maintained by the community that lives there." Though city agency workers are often quick to follow such statements with explanations of how this work helps build community bonds and a sense of connection to the local environment (as Latif did that day), the argument always comes down to what maintenance would cost the agencies. Somehow, the work of this upkeep must lose its value and become volunteer labor, or perhaps become a smaller cost covered by another entity. Sometimes, agency staff suggested that local nonprofits should train, manage, and maybe compensate green infrastructure maintainers, or that the new infrastructures should be designed to require minimal maintenance. In all cases, the goal was to prevent ties between necessary maintenance labor and the financial burden of city employees' salaries and pensions.

For the water managers I interviewed, cost and capacity concerns drove this assumption that the city need not pay for maintenance, a view that reflects decades of neoliberal disinvestment in public infrastructure systems that has reshaped urban provisioning networks across the globe.[24] In contrast to private firms seeking to exploit labor to maximize profit, workers in these public agencies understand themselves to be striving to provide the public with services that their shrunken budgets simply cannot cover. In LA, most are quick to acknowledge the city's ongoing failures in maintaining established infrastructural networks, decay that back in the 2000s, the American Society of Civil Engineers said would cost $30 billion to remediate.[25] The degraded condition of the LADWP's underground network of pipes was made particularly dramatic near the beginning of my long-term fieldwork period in 2014, when a water main break flooded the UCLA campus, causing millions of dollars' worth of damage to university facilities.[26] Given such failures, my interlocutors who worked for the city considered the prospect of both building and maintaining a brand-new network of dispersed, nature-based infrastructure on top of the existing systems utterly beyond the realm of possibility. Unlike the price of land, understood to be governed by the invisible hand of the market, an inescapable fact with which agencies must grapple, the value of certain forms of necessary maintenance labor is seen as far more mutable—work that can be either left undone or excluded from remuneration, and thus the agencies' balance sheets.

Months of conversations with local water managers dealing with storm-water revealed, with some variation, a shared anticipation of a striking trajectory for future waterscape work within LA. These workers described a city to come in which both the human and nonhuman work of storm-water recharge is distributed across the urban landscape, through de-centralized infrastructures and with operations and maintenance (O&M) functions left largely to "the community." Masoud, a bubbly West Asian American engineer in his fifties, was particularly enthusiastic about this topic. When he went to work for the city fresh out of an engineering PhD program in the 1980s, he had been pulled onto big wastewater projects, including major upgrades and expansions to two of the city's sewage treatment plants. But in the early 2000s he shifted his attention to urban runoff. Sanitation had recently established a group focused on stormwa-ter management, and Masoud was keen to get involved. After more than a decade of work on the issue, he was quick to confirm that he saw no technical issues with the pursuit of distributed recharge infrastructure. Funding was another story. But he saw a path forward through a com-bination of strategic public investment green infrastructure construc-tion and local communities "taking some ownership" over the resulting installations.

We had our longest discussion of these ideas early on a Monday morn-ing in his Bureau of Sanitation office, which (in contrast to the usual water-themed paraphernalia, like Edgar's LA Aqueduct map) was deco-rated with a poster extolling the virtues of superfoods. After offering me an energy bar, Masoud settled in for a long chat about his hopes for the future of the city's water system. Echoing his comments from several charrettes we'd both attended, he riffed on his dream of collaborating with NGOs to develop a "green university" to train residents in the finer points of maintaining stormwater-capture installations. With such instruction, he mused, the community could operate and maintain the city's new skein of stormwater infrastructures—leaving his agency to function more as train-ers than as the providers of the long-term labor necessary to keeping those installations functional. When I expressed some surprise at this account of public agency retreat, he told me that, after years of overseeing city budgets and infrastructural project financing, he saw this sort of devolu-tion as inevitable in the process of making LA's water system sustainable.

Ultimately, he suggested, achieving sustainability in LA means "a series of community groups managing their own water."

Though water resource management was very much Masoud's livelihood, he was sanguine about the willingness of residents to take on that role as unpaid work within their yards and neighborhoods. Meanwhile, the burden of these new ecosystem duties went unacknowledged—or glossed as simply unavoidable. As earlier chapters have shown, not all his colleagues anticipated such a stark shift in responsibilities, particularly within the realm of wastewater systems. Nevertheless, many of my agency interlocutors articulated redistributing some degree of water management work beyond public agencies as a necessary element of stormwater recharge and climate adaptation for the city.

Speaking in these terms, my interlocutors echoed discourses frequently associated with PES schemes. As noted earlier, such programs are grounded in the assumption that some people will receive ongoing payments to do some sort of ecological labor (or prevent alternative forms of development on a slice of land) in the name of maximizing certain ecosystem services. In this sense, these projects diverge from the green infrastructure installations being championed for LA, which typically offer no such compensation for the work of maintenance. However, like many PES initiatives, the programs I observed sought to shift land management practices on parcels not intensively managed by or, in some cases, directly controlled by the state.[27] As outlined in chapter 3, LA is not seeking to acquire large swathes of new property for green infrastructure installations, even within the aquifer-connected northeastern Valley. The focus is on more distributed strategies that do not require the purchase of land. City-owned parks, power-line easements, medians, and parkway strips are frequent targets for these projects and figure prominently in the city's stormwater planning documents. While the city technically owns such sites, due to budget constraints many of them receive limited oversight, meaning that managing them as infrastructural nature entails a substantive departure from past maintenance arrangements. And city staff like Latif and Masoud tend to assume that an engaged public will play a large role in the upkeep of that retrofitted public space.

As the existence of domestic-scale recharge projects attests, in the LA context private residential parcels, which cover the majority of the city's landmass, are also often considered key sites of infiltration potential. During my months observing in her NGO's office, Ellen spoke frequently of the ecological promise the city's residential landscapes held. Her organization's neighborhood pilot was an undertaking explicitly intended to take advantage of that opportunity. Funded by grants from several public agencies, including the city's water department, her NGO was one of several working on such residential pilots during my fieldwork period. The coexistence of these publicly funded projects was no coincidence: recent stormwater planning documents clearly assume the mobilization of both the domestic spaces within the urban landscape and Angelenos themselves into new water-management roles. For instance, the LADWP's Stormwater Capture Master Plan's "conservative" scenario calls for 1.4 percent of the city's 11,425 acres of single-family home "opportunity" landscape to be retrofitted to infiltrate stormwater on an annual basis (LADWP 2015, 68). Meanwhile, the plan's "aggressive" scenario assumes a 4.4 percent annual adoption rate on these private properties (68). In a metropolis the size of LA, such figures represent hundreds of residential landscape conversions per year, a striking assumption.

With these retrofits, the space of the home and the yard enter the assemblage of infrastructures tasked with managing the flow of runoff, part of a reappraisal process that Meilinger and Monstadt term "infrastructuring gardens."[28] And while the plan suggests incentives and rebates to encourage such yard conversions, there is no reference to the possibility of such support having the ability to offset the ongoing maintenance for the new water installations. An urban future of more dispersed infrastructure, then, is assumed to be a future of widely distributed, unremunerated upkeep labor—that is, of ecosystem duties. In all these accounts, the labor of locals and the work of the urban terrain are both framed as positive externalities, cheap resources that the state can call on to achieve a desired managerial end. Here, "free" human efforts are being appropriated alongside those that the retrofitted terrain provides. As such, this hybrid labor represents a new genre of water management work, which these institutions desire primarily because of the way it naturalizes and cheapens the labor. Notably, these arrangements also represent cases in which residents are implored to take on the work of infrastructural maintenance as an act of care for the broader LA waterscape and community—highlighting the sense of obligation that the term *ecosystem duties* suggests.

ENVIRONMENTAL ANXIETY AND MORAL
ECOLOGIES OF DISTRIBUTED CAPTURE

Over my months observing the operations at the NGO, I was sometimes asked to help draft promotional materials related to the pilot project. During the flurry of work retrofitting properties, the small organization had fallen behind in the representational labor of making the green infrastructure visible to funders and a broader audience through sleek reports and social media campaigns. I was recruited to assist with these tasks, which I found helpful for learning about how Ellen and her colleagues understood their project and wanted others to interpret it. Translating our sprawling conversations into pithy text touting the benefits of taking a distributed, nature-based approach to stormwater management clarified the advantages many environmentalists see in delegating the "work" of water management to the plants and soils and residents of LA.

Familiar with the funding concerns of her public agency colleagues, Ellen instructed me to frame these documents in terms of cost, emphasizing the long-term O&M savings the city could achieve by incentivizing a program of green infrastructure at the scale of the individual residential parcel, rather than developing larger-scale, city-managed water infrastructures. She often emphasized the fact that such expenses could be eliminated if projects were sited on private land and maintained by their owners. "No one expects someone from the city to fix the rain garden in their front yard," she told me more than once.

While Ellen believed in this argument, for her, talking about the distributed green infrastructure approach's low cost was also a shorthand for expounding on its overwhelming efficacy and importance for adapting the city to climate change. She spoke and wrote in terms of O&M budgets when she believed that her audience was thinking in that language. But outside water agency spaces, she operated on a more sweeping and urgent register. Like many of her environmentalist colleagues, Ellen understood relying on the work of residents rather than the government and using small-scale, nature-based strategies to capture and infiltrate water where it falls as the best and possibly only way to avoid future socio-ecological collapse within LA. Home landscapes and individual residents represented severely underused resources in the fight against environmental catastrophes of water shortage, increased flooding, and urban heat island effect, all expected to worsen under conditions of climate change. Ellen was wholly dedicated to her ambitious vision that her organization's pilot (coupled with relentless advocacy) could spur many more Angelenos— particularly those living in aquifer-connected areas of the northeastern Valley—to rework (and maintain) their own homes and yards.[29]

Such framings signal that in the context of LA's waterscape, some efforts to mobilize unremunerated infrastructural labor are directly connected to deeply held concerns about climate change's local manifestations. Yet despite NGO workers' recognition of the importance of green infrastructure maintenance work for realizing water-related climate adaptation goals, that labor often slipped from view when they presented their projects and strategies to the public.

For instance, observing a daylong conference on local water issues, I watched Hanna, a longtime worker at another environmental NGO, explain to a crowd of policymakers and engineers the need for a shift "from government to governance" in the world of urban water.[30] She expounded on the need for Angelenos to take a more active role in water management, to change the local culture and develop a new water ethic. "We want capturing rainwater on your property to become the norm," she told the crowd. She cited Australia as an example of a comparable country where such a transition had already occurred, flashing photos of backyard rain tanks and swales from a recent trip to Adelaide. During the question-and-answer period, I asked whether she had any concerns about the fairness of shifting the work of water management from the public agencies to the residents. Hanna shook her head. "I see nothing but environmental justice here," she responded, then listed the localized benefits such interventions bring: more vegetation and shade trees, lower temperatures, and perhaps even fewer floods. "Doing this will help make a more equitable distribution of environmental services that are lacking in poor areas," she concluded.

While her concept explicitly demands the devolution of work from the state to residents, that labor is de-emphasized in public presentations through a focus on the array of "natural" services and new benefits associated with the new arrangements. Notably, the nature itself is scripted as ameliorative and cooperative, providing benefits but never overgrowing its planned plots. Put differently, we might say that her framing highlights the ecosystem services that these new infrastructures offer to residents while obscuring the attendant ecosystem duties they entail for the same communities—as well as the capacity for nonhuman nature to subvert a program's plan for remaking a local ecology.[31]

When the work of maintenance enters public discourse around such infrastructures, it typically does so as "stewardship" or "management." A few days after my exchange with Hanna, for instance, I attended a public rainwater harvesting workshop that her colleague Andrea ran at their NGO's headquarters. Nearly a hundred people—who, like the NGO workers hosting us, were mostly white and mostly women—gathered on a Saturday morning to learn the basics of capturing rainwater within our home landscapes. Andrea told us that we all shared a new mission: to become managers of our own "mini watersheds," otherwise known as our yards. She offered nearly three hours of details on the finer points of how to dig swales and secure rain tanks, presenting the work of improving and maintaining one's property in this way as affectively rewarding, meaningful action throughout the session. "We're all worried about climate change," she told the group. "You'll feel better if you just DO something, just one thing to save some water. It doesn't matter if it's small, don't hold yourself to a standard of perfection—but don't let yourself feel stuck." Similar narrations of home landscapes and intimately scaled replumbing labor marked several LADWP-sponsored workshops and presentations that I attended over the course of my fieldwork. While typically delivered by NGO workers and "green" landscapers working on contract for the agency, the events were overseen and administered by city staff, signaling support for the framing.

In such settings, the labor of retrofitting and maintaining an absorbent home landscape becomes a tool for managing space, water, and emotional well-being. Capturing water is presented as an act of care for the environment and the self, comfortably removed from the realm of monetarily valued work. This framing resonates with other ethnographic accounts of embodied ecological labor, as well as with rhetoric deployed in contexts where neoliberal austerity redistributed to religious groups care services that welfare programs once provided.[32] The appropriate reward for work, in such arrangements, is understood as at once moral and emotional, pleasures disassociated from economic valuation. Such characterizations show how these intimately scaled practices can become imbricated in moral ecologies, slotted into understandings of how landscapes, infra-structures, and flows should be arranged for the good of the city and the civic-minded individual.

The settings of their articulation, however, highlight a key dimension of these visions of work within LA's future waterscape: they largely emerge from environmental NGOs and public agencies. As such, they stand in contrast to programs of unremunerated provisioning work that have es-tablished themselves within marginalized groups in other cities. Within the United States, for instance, in the late 1960s the Young Lords emerged from New York City's Puerto Rican community as an activist group work-ing at the intersection of anti-racist organizing and public service delivery. Pairing public protests with volunteer sanitation work, the group mobi-lized barrio residents to provide the trash-picking and street-sweeping ser-vices that the city had long failed to deliver in their neighborhoods.[33] More recently, residents of austerity-stricken Detroit have coordinated among themselves to deliver bottled water to residents in the face of widespread shutoffs and provide basic maintenance labor on abandoned homes and public spaces.[34] Such grassroots self-help programs clearly diverge from those elaborated here, in which technocratic and environmentalist actors seek to incite comparable forms of unpaid labor in the name of shoring up the city's water provision network. While deploying similar discourses of collective care and obligation, here the desire to enroll such free, dispersed work into urban environmental management is rooted in the very institu-tions that many self-help groups seek to critique and transform through their self-provisioning actions.

Persistent challenges with community outreach for stormwater proj-ects in the northeastern Valley suggest some mismatch between resi-dents' environmental priorities and those of the public agency and NGO workers leading such initiatives. Amy, a former NGO staffer and a vet-eran of several infiltration projects in the area, recounted the difficulties of drawing residents to public meetings on the topic. This was largely be-cause the stormwater infiltration projects weren't seen as a threat and so weren't a local priority, she explained. She recounted struggling to drum up attendance for a forum on green infrastructure projects—and then learning that two nights before her event, 250 residents had attended a local public meeting about a local landfill increasing in size. "These people care about the environment, but they're not necessarily going to drag themselves out to hear a presentation about an alternative to the storm drain," she concluded. Zach, a staffer at a northeastern Valley environmental justice organization, struck a similar note in our discussions of green infrastructure and stormwater recharge. His group focused largely on pollution monitoring and cleanup and local park expansion, which he characterized as activities that reflected community priorities. Increasing stormwater capture and other ecological benefits in the area were projects he saw as "ancillary." While quick to note that many northeastern Valley residents took water conservation very seriously in their own homes, he made it clear that, at least in his experience, the long-term security of the city's water supply was not an animating concern for many local residents.[35]

Through outreach programming, installations in the public right of way, and residential retrofits, NGO and public agency workers sought to shift this orientation and draw more northeastern Valley residents into the sustained work of maintaining infrastructural nature. But as my time around the neighborhood pilot project suggested, the reality of this work can be somewhat less romantic and more annoying than dominant framings suggest. The next section addresses the sense among project participants that this unpaid labor is a burden, which underlines that the targeted distribution of these projects within the LA context can serve to perpetuate established environmental inequities within the city.

ON ENVIRONMENTAL WORK AND ENVIRONMENTAL (IN)EQUITY

Talking with participants in the pilot project, including Ariana, whom I mentioned in the introduction, I frequently encountered frustration with the work that living as a watershed "steward" entails. While many expressed support for the goals of the project and appreciation for the retrofits carried out on their homes, such enthusiasm was often tempered with laments about the labor that proper maintenance required. Another example was Nina, a Southeast Asian immigrant mother of three, who worked as a mail carrier before returning home to oversee her children's dinner and evening routines. "I work ten hours a day, six days a week, then help my kids with homework," she told me. "Maintenance is challenging, the weeds keep coming back. Sometimes I'm tired, I forget about it and the weeds get so high it's like a jungle out there!"

As she sat in her living room on a weeknight, that tiredness was palpable. The early fall sun had been particularly intense that day, making her day's work even sweatier and more draining than usual. The appeal of putting in a couple hours of weeding was, she acknowledged, nonexistent under these circumstances. While she assured me that she liked her yard's revamped aesthetic and especially appreciated how removing sections of lawn had lowered her water bill, Nina acknowledged a sense of persistent dread at the idea of venturing into the garden to tame its plants. Such expressions of displeasure with the unruliness of the domestic nature and the hot, dusty, sometimes overwhelming labor of its management evoke other accounts of physically demanding more-than-human care work, which track the coexistence of affective pleasure with bodily discomfort and even danger that can mark such arrangements.[36]

Teresa's narrative of the project highlights similar feelings of ambivalence. A thirty-year-old Latina college student living with her mother and brother, she genuinely delighted in her home's new landscape. Though she loved the lawn they had before, a much more typical landscaping choice in the neighborhood, she much preferred the look of the parkway swales and lavender and sage bushes that had replaced it. The idea of caring for her yard as a form of stewardship also resonated. When we met, she spoke at length about her satisfaction in "helping the earth" and contributing to the city's water supply. But Teresa admitted to struggling with both the hours required and the knowledge necessary for maintaining this yard of "unfamiliar" plants. The lawn had been straightforward, a known quantity; the explosion of growth that crept onto the sidewalk and drew angry calls from her neighbors, less so. She laughed while recalling her harried texts to staffers at the NGO, desperately seeking pruning advice. Overall, Teresa told me, she appreciated the project—but she found it exhausting sometimes.

While acknowledging that a yard will always require some upkeep, nearly all my interviewees affirmed that the green infrastructure demanded considerably more labor than the turf that it replaced. Notably,

Figure 18. Teresa's parkway and yard. Note the plants creeping into the sidewalk, an ongoing problem that had led to neighbors complaining. Photo by author.

while my interviews suggested that women tended to take on more of the maintenance labor, participants discussed the failures to keep up with the work primarily at the household scale, suggesting a sense of collective inability to arrange family life in a way that facilitated adequate tidying and trimming of the infrastructures. Participants' struggle to manage the work became particularly apparent when the NGO wanted to show funders and policymakers around the pilot project to highlight the program's efficacy. Scheduling a tour for an outside group was frequently followed by planning a preemptive day for NGO staff to spend in the neighborhood clearing and pruning rain gardens, scooping accumulated garbage and silt out of parkway swales, and removing piles of leaves from gutters and rain barrel screens, among other tasks. My own experiences sweating through afternoons of this work sometimes included brief, awkward encounters with project participants leaving for or returning from their jobs and either apologizing or sighing in frustration about the yard work they had left undone.

Though participants had signed commitments pledging to complete the maintenance on their own, in practice these were difficult to enforce for an understaffed NGO, and so the organization typically chose to assist with the work rather than threaten consequences. Staffers would also sometimes invite children from participating households to help with this work in neighbors' yards, as discussed at the chapter's outset. In a wealthier area, a more permanent version of this arrangement, in which participants paid a gardener to take on the maintenance, might have emerged. My interviews, however, revealed that most participants did not consider it realistic for them to cover this expense. As such, the moments when the installations were called to perform a communicative and legitimizing function for the NGO served to highlight the challenges of upkeep for the residents.

Among themselves, LA-based NGO employees working on green infrastructure projects are quick to acknowledge the burden of labor that such home landscape transformations entail, and to strategize about how to best support residents in carrying it out. But for some, the experience of participating in this labor raised questions of fairness. Eric, the NGO worker who had championed the "parks and jobs and resilience" that an aggressive program of stormwater recharge infrastructure could offer LA, laid out these worries for me at a coffee shop one Saturday morning. His organization had helped to develop a green infrastructure retrofit project a few years back, reworking the yards and parkway strips of a block in a majority-Latinx, low-income neighborhood in the northeastern Valley. Soon after the completion of the construction, however, elements of the new landscapes began to malfunction. The project team had designed a "curb core" inflow system for the parkway strips, slicing holes beneath the curb rather than making large, open cuts—and the cores were clogging remarkably quickly. Such blockages meant that water from the gutter was not flowing into the parkway swales, and so the green infrastructure was not absorbing nearly the volume of runoff for which it was designed.

The residents struggled to keep up with the work of unstopping the cores, so Eric began to make trips to the project site to assist them. He learned quickly that this was unpleasant labor, requiring him to lie on his belly in the street and stab at the clogged cores with a broom handle. Beyond souring him on the curb core as a design element, the experience

Figure 19. Partially clogged curb core in the northeastern Valley. Photo by author.

had made him question the fairness of placing such projects on private residential parcels. Noting that, as a green infrastructure advocate, he "can't say this publicly in LA right now," he gave me his harsh assessment: "This is a project that effectively makes twenty-four low-income families responsible for the maintenance of infrastructure in a way that no one else in the city is."

For Eric, this realization stung. As he explained, the pursuit of environmental equity for underserved sections of the city was what initially spurred him to work on water issues. On a sunny day over a decade earlier, he had suddenly found himself driving through three feet of water on a road in the northeastern Valley. Though struck by the extent of the flood (remnants of the previous day's downpour), he was not entirely surprised. Enrolled in a geography course at a nearby university, Eric had learned about LA's unevenly distributed flood control infrastructure. Recalling the incident, he explained that experiencing the localized phenomenon he'd read about made the whole arrangement feel tangible and urgent—and helped launch him toward the job with a water-focused local NGO that he held when we met. His own neighborhood didn't suffer from these sorts of inundations, so why should this one? Like many of my other environmentalist interlocutors, he saw nature-based stormwater infrastructures as an ideal way to address such persistent local flooding problems, given their capacity to recharge the basin and improve local water quality in the process. But his recognition of the ongoing human effort necessary to realizing nature's work was pushing him to reconsider the best approach to such installations.

Eric's assessment captures the contradiction at the heart of efforts to site green infrastructure projects on private land and produce ecological benefits at a larger scale—and agency workers' acknowledgment that they anticipate "local communities" will play a substantive role in maintaining infiltration infrastructures sited on slices of public land. While these are conceived as installations for the public benefit that could, in theory, be sited almost anywhere, in practice they will almost certainly not be everywhere. It also reflects a reality of these projects: in the LA context, water agencies and NGOs aim to concentrate these new infrastructures on the aquifer-connected lands where the projects can help direct stormwater underground.

This spatial focus makes good hydrological and engineering sense. But approaching the projects through the lens of labor raises questions regarding the burdens associated with such a distribution. Green infrastructure requires substantial, ongoing inputs of work to provide the city with ecosystem services. Those shouldering the heaviest burden of the work will experience it as an ongoing nuisance, not unlike the disproportionate concentrations of air, soil, or water pollution that typically occupy environmental justice scholars and activists. As many have shown, in the US context in general and the LA context in particular, these environmental harms are typically concentrated in low-income communities of color, suggesting the importance of attending to the ways in which race and class map onto patterns of the distribution of environmental labor.[37] As discussed in previous chapters, in this context the work will be concentrated in racialized, relatively poor, disproportionately polluted neighborhoods, reinscribing existing patterns of environmental inequity. Green infrastructure installations may indeed bring localized environmental benefits to these areas, but by twinning such amenities with unwaged work, in this case they reproduce established patterns of environmental burden and the devaluation of marginalized communities' labor.

My purpose in highlighting the unequal distribution of this form of adaptation work—and the ways in which it mirrors other patterns of environmental injustice—is not to suggest that there should be no green infrastructure installed in these neighborhoods, or that such lopsided dynamics are inevitable. Rather, it is to show how tracking the distribution of such work can illuminate emergent arrangements of environmental burden. In this context, policymakers and advocates understand distributed recharge projects as an approach to water supply augmentation that addresses citywide environmental problems quickly and cheaply, essential characteristics in an era of ecological crisis and strained municipal budgets. If these goals are valuable, then addressing the unequal, uncompensated burden of labor that their presence could bring to marginalized communities becomes even more pressing.

Developing the notion of ecosystem duties, terminology that places the work of environmental maintenance on the same plane as the benefits of ecosystem services, pushes considerations of such arrangements to fully weigh the associated burden of human labor and to confront its articulation with other forms of ecological harm. Drawing attention to the nonvaluation of this work also raises key questions about how such exertions might be reconceived as labor that could direct new resources directly to residents and community groups of the northeastern Valley. As the PES literature suggests, infrastructural nature projects can serve as conduits for redistributing wealth to marginalized communities.[38] Reframing green infrastructure maintenance as labor that could be linked to livelihoods has the potential to open a broader conversation about how the diverse forms of work that climate adaptation entails should be distributed and valued, as well as the infrastructural and funding paradigms challenged in the process. Mobilizing the labor of nature in such arrangements requires long-term human work; grappling with this reality, I suggest, offers the potential for such projects to transform more than just systems of ecological flows.

Some of my NGO interlocutors raised similar concepts during my time in the field. The idea that the work of both building and sustaining infrastructural nature could be a source of well-paid "green" jobs was a frequent topic of discussion. Ellen was particularly invested in this notion, understanding such forms of employment as a key complement to volunteer exertions of northeastern Valley residents. She worked doggedly to interest collaborators based in city agencies in the idea of establishing a dedicated training and certification program for landscapers who might want to offer green infrastructure-related services to clients. While frequently rebuffed, she persisted, driven by hope that one day both rain and the work of capturing it for the city might enable residents to sustain their lives in LA.

To remain functional, infrastructural nature demands a long-term collaboration between human actors and other-than-human natures. Approaching these efforts as labor foregrounds the issues of equity that the language of "ecosystem" or "nature's" services obscures. Using the language of ecosystem duties to signal the human work that must accompany that of ecosystem services offers a helpful way to signal the extent to which these infrastructures bring both benefits and labor to the landscapes they are placed within.

As several of my interlocutors described, the benefits such installations provide can exceed the realm of ecology: maintaining these infrastructures is work that can also bring pleasure, satisfaction, and a deeper sense of connection to local ecologies and communities. This is hybrid labor that can produce novel multispecies relations and sometimes political formations. Considering the ambivalent character of such labor in this context highlights the salience of power-laden distributive questions around its mobilization. If the provision of ecosystem services like stormwater recharge entails work for both a landscape and some people, we must attend to the questions of *which* landscapes and *which* people are tasked with the labor, and which ones are receiving the benefits. A mismatch between the distribution of the services and the duties is helpful for assessing the role such infrastructures play in reproducing patterns of environmental injustice—particularly when that work is unattached to a wage.

The configurations of hybrid labor I encountered in LA's northeastern Valley suggest one potential outcome: that when urban terrain is reworked as a landscape of green infrastructure, new forms of inequality can emerge as selected landscapes within the city are managed to provide ecosystem services for the entire city. In their attempts to replumb LA in the context of limited public funding, city water managers and environmental advocates seek to mobilize cheap work from the northeastern Valley's landscape and residents. Given the city's protracted failures to provide adequate public services to these communities, the demand for and non-valuation of their maintenance labor can be read as an arrangement that reinscribes established patterns of spatial inequity within the city. Highlighting these concentrated labor burdens signals the political nature of efforts to redistribute the work of urban water management—particularly in cases where "nature" is scripted to handle the bulk of that labor.

Epilogue

California's winter of 2023 was wet—by some metrics, historically so. Storm after storm rolled in from the Pacific, blanketing the Sierra Nevada with a record-breaking snowpack. Closer to sea level, downpours and high-elevation snowmelt swelled rivers, leading to floods, evacuations, and the reemergence of a long-dry lake within the state's agricultural interior. While the damage was less dramatic in urban Southern California, the season brought localized flooding and repeated freeway closures within and beyond LA.

The rain also sparked despondent accounts of the city's failure to capture more of the runoff rolling through city streets and flood control channels for water supply purposes. "Despite heaps of water pouring into the area, drought-weary Los Angeles won't be able to save even half of it," an *LA Times* journalist noted that January. "The region's system of engineered waterways is designed to whisk L.A.'s stormwater out to sea—a strategy intended to reduce flooding that nonetheless sacrifices countless precious gallons."[1]

Given the string of dry years preceding that soggy one, such frustrated characterizations of the city's runoff likely resonated with local readers. Not long after the 2012–2017 drought that colored my years of fieldwork for this project, California crept into another protracted period of aridity. As the COVID-19 pandemic dragged on, familiar images of shrunken reservoirs began to surface on the evening news, just as they had throughout the previous dry spell. But it wasn't all reprise. During the drought of the early 2020s, some critical nodes within the state's water provision network were more desiccated than ever previously recorded. Of particular note was the Edward Hyatt Powerplant, a hydropower generation facility at Northern California's Oroville Dam. Due to unprecedently low water-storage levels in its reservoir near the end of 2021, the facility was forced to shut down operations for the first time in the plant's fifty-three-year operating history. As such events suggest (and as climate modelers have been anticipating for years), lately California's characteristic feast-famine hydrology has been looking more and more like a caricature of itself, swinging between far too dry for far too long and dangerously soaked. Under these conditions, the spatial logic of LA's water provision system, reliant on flows piped in from the US West's arid interior while efficiently siphoning local rain and wastewater to the ocean, appears increasingly puzzling and precarious.

The image of a disheartened reporter watching LA's stormwater slide toward the sea is in many ways a discordant echo of a more famous scene from the city's water history. On November 5, 1913, flows from its newly completed aqueduct first entered LA's water system. The event was staged as a grand public ceremony, for which an estimated thirty thousand residents gathered. When the sluice gates opened, allowing a torrent of Owens Valley water to thunder down the pipeline's open-air terminus in the northeastern San Fernando Valley for the first time, city water chief William Mulholland famously instructed the crowd, "There it is, take it." And so they did, tapping into the city's well-stocked water grid to develop subdivisions and factories and swimming pools and lush gardens, spreading the aqueduct's water across LA's urbanizing landscape. Drawing on ever-more-distant sources, the agencies tasked with sustaining flows within that network have kept water deliveries reliable through the droughts and floods and economic booms and recessions and uprisings and pandemics of the following decades.

As detailed throughout these pages, replumbing LA through projects of in-city wastewater reuse and stormwater recharge is premised on maintaining such steady provision arrangements. A desire to sustain the city's approximation of the modern infrastructural ideal of unceasing net-worked flows underpins much of the work I've recounted here, buttress-ing other forms of water system stasis in the process. Crucially, this goal guides public agency efforts to protect claims to distant water sources, in addition to attempts to produce new water supplies within LA's bor-ders. The LADWP's dismissive response to environmentalist demands to halt diversions from the source waters of languishing Mono Lake during the early 2020s drought underlined that orientation, demonstrating the agency's determination to sustain its material links to such distant land-scapes).[2] While the recognition of climate change has undermined the sense of resource security once associated with these distant waters, it has not severed ties to them. Their endurance reflects the retrenchment and revitalization of big dam and urban-rural transfer projects documented across the globe in recent years.[3] Critiques of the socio-ecological sustain-ability of such arrangements may be prominent, but the pipelines largely persist.

However, at many moments, environmentalists and water managers alike present the "taking" of in-city resources as a shift with the potential to do far more than enhance resource security in water-stressed LA. In-frastructures of reuse and recharge are often scripted as installations with the potential to bring a range of ecological benefits to the urban fabric, enhancing greenery and carbon sequestration capacity and water qual-ity in local creeks in underserved corners of the city. Some also suggest that by drawing residents into the work of water provision through their maintenance, distributed infrastructural nature can seed new relations of ecological care and political mobilization, providing the basis for new urban environmental relations and governance arrangements. Such as-pirations reflect the capacious sense of possibility frequently articulated in connection with the forms of material transformation that climate ad-aptation entails, particularly the hope that—if done well—this transition could address a range of historical harms and broken systems within the social fabric.

But the torrents of stormwater barreling toward the Santa Monica Bay in the soggy days of early 2023 emphasize that transforming LA's in-city water flows into municipal resources is a project likely to remain in progress for the foreseeable future. Despite the 2018 passage of Measure W, a parcel tax that allocates $280 million annually for stormwater infiltration across LA County, progress toward expanding recharge has been slow, with many projects mired in the planning and design phases.[4] Meanwhile, wastewater remains a potable water resource of the future within the city of LA. Thanks in part to a low-interest $225 million loan secured through the EPA's Water Infrastructure Finance and Innovation Act, in early 2023 the city contracted outside consultants to design and build advanced water purification infrastructure at the Donald C. Tillman Reclamation Plant in the San Fernando Valley. If the agency's optimistic timeline holds, the cleansed effluent will be piped to the Hansen Spreading Grounds and stored in the San Fernando Groundwater Basin by the end of 2027—a full twenty-seven years after LA's last, ill-fated sewage-to-drinking-water project was shuttered.[5]

Such developments suggest that some of the local frictions regarding the price of recharge infrastructure and acceptability of reuse facilities that the previous chapters documented may have waned somewhat in recent years. The establishment of new, dedicated funding sources at multiple scales for such projects has likely enabled this trajectory. However, the slow pace of infrastructural change within the urban landscape underlines the diffuse, wide-ranging forms of work—including bureaucratic, design, and outreach labor—necessary to rerouting urban flows. In contrast to the temporary work camps established along the path of the LA Aqueduct to facilitate that pipeline's rapid construction at the beginning of the last century, much of this labor will take place in air-conditioned offices dispersed across (and sometimes beyond) Greater LA. While often distant from the prospective reuse and recharge infrastructures, such work is nonetheless essential to their development.

Likewise, distributed forms of maintenance labor will be necessary to sustain these facilities when construction is finally complete. Perhaps unsurprisingly, given the sluggish pace of infrastructure development, media reports and planning documents rarely discuss the status and value of such long-term work—suggesting that the arrangements depicted within these pages likely persist. Paid agency staff will maintain many centralized infrastructures (like wastewater recycling plants and upgraded spreading-grounds facilities), while the upkeep of distributed infrastructural nature installations (like infiltration basins in a power line easement or a front yard rain garden) will probably fall to unremunerated residents or volunteers from community groups within the northeastern San Fernando Valley. While other-than-human nature will unquestionably play a role in those recharge arrangements, that hybrid labor will involve substantive human exertion as well. In this context, the result will be an unevenly distributed set of ecosystem duties, unpaid work mobilized primarily from marginalized communities in the name of securing a collective resource base for the city. As I have illustrated, such labor can come with pleasure, seed new multispecies relations, and spark new forms of political engagement. But this practice is also undeniably free work, drawn primarily from a landscape long marked by environmental disorder and infrastructural underinvestment from the state—an arrangement that signals the broader relevance of ecological labor for theorizations of environmental justice, particularly in the context of a changing climate.

In many settings, productive work and environmentalism have come to be understood as inherently conflictual projects. Drawn from a bumper sticker, the title of historian Richard White's classic account of fights over logging in the US Pacific Northwest captures the popular sense of such divides: "Are You an Environmentalist or Do You Work for a Living?"[6] Of course, as White aptly notes, such a sense of nature-labor dichotomy is best approached as an artifact of a particular historical moment and set of political economic arrangements, given that working with the material stuff of the environment has long been the primary way that most people engage with nature.[7] Ethnographic accounts of such work also demonstrate the peculiarity of equating labor-in-nature so stably with resource extraction and landscape degradation. In particular, the fields of environmental anthropology and political ecology have long foregrounded forms of ecological labor (often vilified by "expert" colonial, state, and NGO actors) that Indigenous communities have undertaken to regenerate ecological functionality.[8]

While recognizing the extended histories of such work (alongside their frequent problematization), we do well to acknowledge that recent decades have seen a marked expansion of state-, donor-, and investor-driven projects of ecological restoration and regeneration, signaling new forms of imbrication between environmental labor and monetary valuation.

Underpinned by the ascendance of the ecosystem services paradigm, large-scale programs of tree, biomass, and soil cultivation; wetland and stream restoration; and targeted forest thinning and controlled burns all mobilize diverse forms of human work in the name of enhancing land and waterscape functionality. Increasingly, we live in a world where mobilizing selected functions from the environment is understood as a pressing global undertaking, with particularly high stakes for the project of climate adaptation.

But how, exactly, do these efforts relate to the need to, as the bumper sticker puts it, "work for a living"? As a growing literature attests, answers can vary wildly. Sometimes the labor is mobilized for free or for (too) little money (to live on), and sometimes it offers a pathway for residents to sustain their homes and communities. Such variation signals the ambiguous, largely unsettled relationship between such exertions and processes of valuation and the attendant potential for this work to become a potent site of claims making for marginalized communities—particularly given the state's central role in so many of these projects. The prospect of transforming such unevenly distributed ecosystem duties into livelihoods hints at a deeper, more just rearrangement of urban land and waterscapes that a program of replumbing the city could entail in a place like LA—as programs of urban ecosystem services enhancement could in cities across the globe. As proponents rightly note, the infrastructural nature installations that enable stormwater infiltration for public supply have the potential to bring an extensive range of local ecosystem services to landscapes long marked by racialized forms of pollution and infrastructural neglect. Valuing and funding the human labor that sustains such ecological functions could serve as a mechanism to redistribute wealth alongside these benefits.

Such possibilities signal how approaching environmental transformation through the analytic of adaptation work, as I've done throughout this account of the LA waterscape, might provide traction for a new range of environmental justice questions and demands within the urban fabric, particularly in the context of climate change. As I've shown in this case, the devolution of water management labor and infrastructure is in many ways an ambivalent, contradictory process. State institutions valorize some genres of spatially distributed water-directing installations and work while discouraging others. Further, beyond rerouting material flows, some see such arrangements as potential on-ramps for new forms of political mobilization within the urban environment. Recognizing the possibility for ecological labor to produce new burdens, pleasures, paid employment opportunities, and oppositional engagements with the state signals that programs and sites intended to incite such practices could play a key role in reconfiguring the urban socio-ecological fabric. Holding the potential to exacerbate or redress established patterns of environmental injustice, such arrangements deserve careful attention, particularly when mobilized for the sake of adapting "the city" as an undifferentiated whole.

Notes

INTRODUCTION

1. Per my research protocol, throughout this manuscript I use pseudonyms to refer to my interlocutors unless otherwise noted. In some cases, identifying details have also been adjusted or obscured for the sake of preserving anonymity.

2. A quick disambiguation note: two California valleys play substantive roles in the LA waterscape, and in this book. The San Fernando Valley, referenced here, sits mostly within the city of LA and above the subsurface San Fernando Groundwater Basin. The Owens Valley, a major source of LA's water supply, is somewhat farther afield, located roughly two hundred miles northeast of the city on the eastern side of the Sierra Nevada Mountains. In extended discussions of the San Fernando Valley, I follow colloquial LA practice and simply call it "the Valley" or, in more specific cases, "the northeastern Valley" after an initial identification. In contrast, whenever I refer to the Owens Valley, I use that valley's full name.

3. The San Fernando Basin, the Sylmar Basin, the Verdugo Basin, and the Eagle Rock Basin are four separate aquifers located within the San Fernando Valley. Though the four basins together are sometimes referred to as the "San Fernando Valley Basin," and all are overseen by the Upper Los Angeles River Area Watermaster, my discussion here—and elsewhere in the text, unless otherwise noted—focuses only on the San Fernando Basin.

4. As in such framings—and reflecting a rich vein of geographical literature on the concept of landscape (see Mitchell 2002; Malpas 2011; Stilgoe

2018)—throughout the book I use this term to connote configurations that are actively produced on both material and semiotic registers. Embodied human labor and representational practices are understood as constitutive, entangled elements of this process. As Don Mitchell, a consummate chronicler of such relationships within the California context, puts it, I see the "work and the 'exercise of imagination' that makes its products knowable" as underpinning the production of the landscapes discussed here. See Mitchell (1996, 1–2).

5. Aquifers are far more materially complex assemblages than simple storage tanks, marked by multiple, ongoing forms of seepage and flow. But while my interlocutors understand this complexity, they let the notion of these basins as sites to stockpile water guide their approach to LA's terrain. As such, while I dwell on the particulars of aquifer seepage and flow in select moments, I foreground the basin's imagined "storage tank" function throughout the book, exploring how this notion organizes attention, work, and flows of resources within the city.

6. Throughout the text I focus primarily on the water supply, infrastructure, and politics of the city of LA—an entity I typically refer to simply as "LA." However, it bears noting that the city (with its population of roughly four million) is situated within LA County (with a total population that has hovered around ten million in recent years). While the city contains and manages much of the relevant infrastructure and water discussed throughout the book, county-, state-, regional-, and federal-scale entities own and oversee some others. These distinctions are elaborated in further detail later in the text.

7. Today the LA Aqueduct system stretches 338 miles, thanks to a 1941 extension that expanded the pipeline north to Mono Lake, the site of long-term legal headaches for the LADWP.

8. It bears noting that sections of LA County beyond LA city limits are plagued by dangerously poor drinking water quality. Some residents within the city also lack stable access to clean potable water, conditions I elaborate in chapter 1.

9. LADWP (2021).

10. As discussed further later, while climate change is anticipated to stress this system water provision, the network's vulnerability is also rooted in past infrastructural development and water governance decisions. See Pincetl et al. (2019).

11. In the name of local flood protection, the United States Army Corps of Engineers encased most of the fifty-two-mile-long LA River in concrete in the 1940s. Scholars have analyzed the self-evidently man-made nature of the watercourse—and the ongoing environmentalist efforts to "renaturalize" it—in detail in several critical works since the turn of the century. See Gumprecht (2001), Desfor and Keil (2004), Orsi (2004), and Gandy (2014). Here, I focus on flood management infrastructure beyond the river channel, a decision guided by the recognition that much of the current work related to the LA stormwater that eventually flows into the river targets land well removed from its channel.

12. Von Schnitzler (2013, 673).

13. Fredericks (2018, 15). See also Nucho (2017), Anand et al. (2018), and Hetherington (2019).

14. See Anand (2017), Chalfin (2023). See also Latour (2004) and Bennett (2010).

15. Nelson and Bigger (2022, 87; emphasis in original).

16. See Carse (2012, 2014), Lock (2023), Nelson et al. (2020), and Wakefield (2020). See also Zee (2021), Hoag (2022), and Kurtiç (2023).

17. Unlike inherently exhaustible coal mines or oil wells, these assemblages are presented as capable of providing perpetual services if managed properly. Notably, however, interventions within the category of infrastructural nature diverge from one another on a different temporal register. Some of these schemes seek to preserve ecosystem functions deemed valuable, resembling older efforts oriented toward protecting "wilderness" (see Cronon 1996). But many others, like the LA water projects I examine here, are largely future oriented, intended to augment or develop or restore these capacities within selected landscapes.

18. Throughout the book, when the category of the human is invoked to mark a distinction from beings and materials typically coded as "nature," I do so with the recognition that humanness itself is a contested category, too often synonymous with that of white, Western man. Work performed by those beyond the normative white, male category is frequently occluded or naturalized. See Weheliye (2014) and Wynter (2003). With this in mind, I attend carefully to how race, class, and gender articulate with theories of labor broadly and with the work explored in my ethnographic context.

19. See Barnes (2014), Björkman (2015), Von Schnitzler (2016), Anand (2017), Butt (2023), and Zhang (2024).

20. Readers may note some resonance between such arrangements and urban scholar AbdouMaliq Simone's influential concept of *people as infrastructure*, which extends "the notion of infrastructure directly to people's activities in the city" and conceptualizes such activities as "a platform providing for and reproducing life in the city" (Simone 2004, 407–8). Developing this concept through an ethnography of inner-city Johannesburg, Simone foregrounds the bottom-up, provisional nature of these efforts, a notable contrast to cases like LA's, in which state agencies and NGOs seek to incorporate such dispersed efforts into their infrastructural arrangements.

21. Battistoni (2017, 6).

22. Also building on Battistoni's framework, geographer Marion Ernwein (2020) adopts a similar framework in her analysis of work arrangements associated with urban greening in Switzerland, points that are further articulated in the coauthored introduction to the volume *How Plants Work* (Erwein, Ginn, and Palmer 2021).

23. Besky and Blanchette (2019, 5). See also Daggett (2019).

24. See Gandy (2022) and Stoetzer (2022).

25. See Pulido (2000).

26. See Maida (2011) and Shamasunder et al. (2015).

27. See Johnson et al. (2023) and Mills-Novoa (2023).

28. Scholars have also explored and problematized such narratives on a range of registers. See Erie (2006), Deverell and Sitton (2013), and Sizek (2023).

29. Both tribes have contested city and county approaches to local water and land management within LA's boundaries, particularly those related to the LA River. As of November 2024, neither group is federally recognized, complicating their ongoing efforts to pursue Land Back and pressing ecological priorities. During my fieldwork period, tribal representatives were rarely present in the technocratic spaces in which I conducted participant observation. Accordingly, their perspectives are not centered in my accounts of these settings, a gap that reflects and to some degree reproduces the marginalization of these communities within LA water politics and management. I recognize this omission of Indigenous voices and priorities in my work as a weakness, one in a long line of similar failures by settler writers. Tongva scholar AnMarie Medoza's research on tribal marginalization and resistance during the most recent LA River planning process will, when published, offer a valuable corrective to such scholarship.

30. Reisner ([1986] 1993, 53).

31. Before the construction of the first LA Aqueduct, the city relied exclusively on local surface and groundwater for its supplies. In the late eighteenth century, Spanish colonizers constructed a network of irrigation ditches for their growing settlement, organized around the Zanja Madre (mother ditch) that ran from the LA River to the town plaza. A standing committee of the Pueblo government supervised the management of the system. Following a brief stint as part of Mexico and the 1848 Mexican–American War contesting that acquisition, California was admitted to the United States in 1850. Several private water companies emerged in the years that followed, consolidating as the Los Angeles City Water Company and overseeing the development of the city's domestic waterworks over the decades that followed. Following protracted negotiations, the city assumed public control of that system in 1902, delegating control to its newly formed LA Water Department. The reacquisition was essential to the city's plans for raising money to build the aqueduct through municipal bonds. The public voted overwhelmingly to approve the bonds in 1906, and construction on the aqueduct commenced the following year. The Water Department was renamed the Bureau of Water Works and Supply, then merged with the Bureau of Power and Light to become the LADWP in 1937.

32. The city hired thousands of workers to construct the aqueduct, housing and feeding most in temporary encampments along its isolated path. While the labor was compensated, this was by no means an idyllic arrangement. The work camps were marked by strikes and conflicts regarding compensation and food

prices. See Piper (2006). Further, access to these jobs was circumscribed by the de facto and de jure racist hiring practices characteristic of LA's municipal employment practices during that period, severely limiting the number of Black, Asian, and Native workers within the project. See Nadeau (1960) and Van Bueren (2002).

33. See Hughes, Pincetl, and Boone (2013).

34. See Diffenbaugh, Swain, and Touma (2015).

35. See Huang, Stevenson, and Hall (2020).

36. See LADWP (2021, ES-19). These figures were derived from the graphs printed on the report's indicated page by multiplying the listed total volume of water by the percentage attributed to each category.

37. See Butler (1993) and Davis (1998).

38. See Randle (2022b).

39. See Bacigalupi (2015) and Watkins (2015).

40. Günel (2016, 291).

41. Barnes (2016), O'Reilly (2016), Knox (2020), and Petryna (2022).

42. Technical expertise on how to adapt landscapes to the predicted stressors remains highly incomplete and is often shaped by the priorities of extra-local funding agencies and other elites—as demonstrated by a rich and growing body of ethnographic scholarship. See Cons (2018), Paprocki (2021), and Vaughn (2022).

43. See Günel (2016, 2019).

44. See Tsing (2015).

45. See Hart (1996), Mendoza (2019), Bertenthal (2021), and Borgias (2024a).

46. Analyzing such arrangements in the context of the Southern California waterscape, geographer Alida Cantor (2021) helpfully glosses such remote landscapes as *hydrosocial hinterlands* to signal their entanglement with urban sites of consumption.

47. See Heynen, Kaika, and Swyngedouw (2006).

48. See Needham (2014), Hommes and Boelens (2017), Powell (2018), and Saguin (2022).

49. Nelson and Bigger (2022, 91).

50. See Hays (1959) and McGee (1909).

51. While the rise of the ecosystem services paradigm is detailed later, this book does not treat the term *ecosystem* as a central object of analysis. Though cognizant of the strand of scholarship that places the rise of the ecosystem concept within the broader ascendance of systems thinking in the post–World War II era (see Olson 2018), I follow in this context my interlocutors in the policy and activist realms, who tended to use the phrases "nature's services" and "ecosystem services" interchangeably in discussions of the green infrastructure installations analyzed here.

52. See Gómez-Baggethun et al. (2010).

53. See Nelson (2015).

54. See Costanza et al. (1997).

55. See Goh (2021) and Tozer et al. (2023).

56. See Williams (1975) and Cronon (1996).

57. See Rademacher (2011) and Rademacher and Sivaramakrishnan (2013).

58. See Cousins (2017a) and Meilinger and Monstadt (2023).

59. Scaramelli (2021, 10). See also Scaramelli (2019).

60. See Cerra (2017), Willems et al. (2020), Riedman (2021), and Lamond and Everett (2023).

61. Zhang (2020, 96).

62. See Stoetzer (2022) and Barua (2023).

63. See Fisher, Svensden, and Connolly (2015), Boyer (2024), and Maurer (2024).

64. Such formulations mirror critical scholars' accounts of the subject formation through the work of environmental care and stewardship. Some strands of this scholarship are grounded in Foucauldian notions of power and highlight how such labor can facilitate the internalization of state-driven norms and narratives. See Agrawal (2005), Singh (2013), and Radonic (2019b). Others foreground the more-than-human ethics that can emerge through these entanglements, emphasizing their ambivalent character. See De la Bellacasa (2017) and Parreñas (2018).

65. Karen Piper offers a particularly robust account of these challenges in her book *Left in the Dust* (2006), recounting a tortuous struggle to access LADWP archives and interview department personnel. While I do not dwell on the challenges of access within the body of the manuscript, her narrative certainly resonated with my experiences.

66. In addition to the challenges related to secrecy and power outlined in classic anthropological texts on "studying up" (see Nader 1974; Gusterson 1996), I also learned to navigate those associated with email gatekeeping, as deftly analyzed by Daniel Souleles in his accounts of research among finance workers in New York City in the 2010s. See Souleles (2018, 2021).

67. I also conducted interviews with six low-water garden designers, one of whom allowed me to accompany her on a handful of yard consultations via an LADWP outreach program. While I do not discuss this material at length within the book, the experiences shaped the analysis of home-scale water management presented in chapters 2 and 5.

1. PUBLIC AGENCY WORK

1. This field trip's timing, scheduled during work hours on a weekday, hints at a dynamic that marked many of the water-related public forums and outreach programs I observed during my time in LA. While often free and open to anyone

who registered online in advance, events were frequently held at times that made it exceedingly difficult for anyone working regular hours in jobs unrelated to the topic to attend. As a result, the crowds at these events were often dominated by retirees and NGO staffers.

2. Notably, in the case of LA's plant closure, one of the key instigators of the public pushback to the water source acknowledged to me in a 2015 interview that even at the time he was more concerned about the prospect that expanding supply in this manner would raise his water rates than about the safety of the effluent. But as he explained, stirring up disgust at the prospect of drinking toilet water was clearly the more effective way to draw critical attention to the project.

3. See Chahim (2022) and Krause (2013, 2022a, 2022b). See also Lefebvre (2004) and Addie (2022).

4. See Ferguson (1999) and Kaika and Swyngedouw (2002).

5. Ethnographic scholarship has shown that urban water grids marked by unsteady or punctuated patterns of water delivery create heavy domestic-scale water management labor burdens, often in highly gendered ways, and sometimes serve as the basis for critical engagements with municipal utilities. See Bjorkman (2014), O'Leary (2016), Anand (2017), and Truelove (2021).

6. See Ferry and Limbert (2008), Limbert (2010), Mathews and Barnes (2016), Ballestero (2019a), and Stamatopoulou-Robbins (2020).

7. See Haughton (1998), Kaika (2005), Millington and Scheba (2020), among many others.

8. California DWR (2021, 7-1).

9. See Zetland (2009) and Cousins (2017c) for examples of social science accounts adopting this frame.

10. Björkman (2018, 289).

11. Graham and Marvin (2001). See also Melosi (2000) and Gandy (2004).

12. See Kaika (2006).

13. While this characterization fits the city of LA, other sections of urban Southern California are served by private water companies. See Reibel, Glickfeld, and Roquemore (2021). Further, a growing number of public sector water management institutions within the region rely heavily on contract labor, particularly engineering consultants.

14. Such framings also resonate with work from the field of critical disaster studies, which explores the local and regional socio-ecological processes that a trenchant focus on climate change can obscure. Gregory Simon's (2017) writing on California's "incendiary" (rather than "flammable") landscape of wildfire risk exemplifies this tendency, highlighting the long-term political economic processes that have produced such hazardous conditions and that increasingly slip from center stage in discourse on the state's fire regime. While cognizant that framing projects of resource governance as climate adaptation can displace attention from other salient dynamics, throughout the text I foreground actors

articulating this position so I can explore the contours of this contested category in the local context.

15. See Pompeii (2020), Egge and Ajibade (2021), and Méndez-Barrientos et al. (2023).

16. Resilience began its life as an ecological concept in the early 1970s, through the work of C. S. Holling. Influenced by ideas from the emergent complexity science, Holling defined the concept as "a measure of the ability of these systems to absorb changes of state variables, driving variables, and parameters, and still persist" (1973, 17). While critical genealogies of the concept's evolution and spread in the subsequent decades vary considerably, all foreground the growing emphasis on shocks, threats, and disruptions that the paradigm assumes. See Walker and Cooper (2011), Evans and Reid (2014), Watts (2015), and Grove (2018). My case reiterates the salience of Stephen Collier and Andrew Lakoff's account of resilience as intertwined with the "vital systems security" paradigm that emerged largely via the national security apparatus in the post–World War II United States. See Collier and Lakoff (2015, 2021).

17. The COVID-19 pandemic exacerbated these conditions of precarious water access among communities across LA. See Gonzalez et al. (2021). While the ongoing catastrophe of the pandemic began after the fieldwork for this project concluded, it bears noting to highlight that societal disruptions seemingly unrelated to water supply can undermine residents' access to the vital resource.

18. See Meehan, Jurjevich, et al. 2021.

19. See Los Angeles Homeless Services Authority (2022) for results of a recent survey seeking to measure LA's homeless population. Just beyond the city limits, sections of LA County are plagued with dangerously poor drinking water quality. Research shows that such problems are concentrated in water systems that serve low-income communities of color in southern LA County. See Reibel, Glickfield, and Roquemore (2021). Racial capitalism has played an enduring role in structuring this unequal regional network of water systems and must be acknowledged, even when the analysis centers on a relatively privileged jurisdiction (and safe water system) within this urban region.

20. See Meehan, Jepson, et al. (2020). While less emphasized in the critical literature on the global North, similar dynamics frequently mark centralized sewerage networks. Chapter 3 explores the limits of LA's twentieth-century sewer system development and the ecological consequences.

21. I am not arguing that these interlocutors were unaware of or uninterested in local water-access challenges. Rather, my conversations with these workers suggest that issues of plumbing poverty were—at most—peripheral to their understanding of systemic water reliability. I highlight this point to be precise regarding a notion central to their work. As other scholars have noted, while there is general recognition that reliability is a key dimension of water supply, "there has been little consensus on how it should be defined, or consequently

measured" (Majuru, Suhrcke, and Hunter 2016, 2). Such observations point to the importance of carefully considering how a local community of practice deploys the term. It also bears noting that most of the individuals I spent time among worked in roles focused on water supply planning and development. The discussions would likely have sounded different if I had spoken with workers more directly engaged with maintaining the city's water distribution infrastructure, as other ethnographers of infrastructure have done. See Anand (2017) and De Coss-Corzo (2020).

22. As such comments suggest, seawater desalination was not under consideration as a supply option for the city of LA during my fieldwork period. Neighboring jurisdictions, however, were exploring the resource at that time and have continued to pursue it in the years since, in the face of considerable community and regulatory pushback. See Morgan (2020), O'Neill (2023), Williams (2018a, 2018b).

23. See Amironesei and Scoville (2019).

24. See Blomquist (1992).

25. See LADWP (2021).

26. In addition to its rights in the San Fernando Groundwater Basin, the LADWP also has the right to pump limited quantities of groundwater in four other LA County basins: Sylmar Basin, Eagle Rock Basin, West Coast Basin, and Central Basin. As the city's pumping rights within the San Fernando Basin dwarf the volumes it can extract from all the others, it is the primary focus of this book's account. As illustrated by figure 7 in chapter 2, most of these basins underlie jurisdictions beyond the borders of the city of LA. Further, at the time of this writing the LADWP has no active pumping wells in the Eagle Rock and West Coast Basins.

27. Andrea Ballestero's (2019b, 2019c) ethnographic accounts of efforts to "know" aquifers are invaluable for anyone trying to get a handle on these complex, pluri-temporal assemblages.

28. As my water manager interlocutors not employed by the City of LA often pointed out in our interviews, the jurisdiction is a notable water reuse laggard within Southern California. Not far beyond city limits, the LA County Sanitation Districts began using treated wastewater for groundwater recharge back in 1962 at their Whittier Narrows Water Reclamation Plant. And in nearby Orange County, a treatment plant called Water Factory 21 fulfilled a similar function from 1975 until 2004, when it was demolished to make way for the bigger Groundwater Replenishment System facility. See Kiparsky et al. (2021). Like LA, the City of San Diego bowed to protests and halted plans for a large-scale potable reuse facility. After years of struggling to revive the project, San Diego finally broke ground on its plant in the summer of 2021—outpacing LA.

29. See, for instance analyses of oil speculation, copper extraction, and unconventional fossil fuel production in other contexts; see Weszkalnys (2015),

Kneas (2020), and Kama (2020, 2021). Such analyses have served as correctives to narratives that present forms of extraction as inevitable, highlighting the uncertain, processual nature of resource becoming. See Ferry and Limbert (2008) and Kneas (2018). Reflecting on such trajectories, Tanya Richardson and Gisa Weszkalnys suggest that they are best approached not as "a linear unfolding but rather an oscillation between different states of being" (2014, 15).

30. Ormerod (2016, 541). See also Christen (2005), Schmidt (2008), and Duong and Saphores (2015).

31. See Shreenath (2023) and Chalfin (2023).

32. See Nevarez (1996) and Kaika (2003).

33. Given the public nature of Sam's film project and her permission, I use her real name rather than a pseudonym.

34. I highly recommend the *The Longest Straw*, a rich, thoughtful documentary that can be streamed on several online platforms. I analyze the film in detail elsewhere. See Randle (2022b).

35. See Hewitt (2013).

36. See Borgias (2024b).

37. See LADWP (2021, ES-19).

38. See Williams (2018a, 2018b).

39. See Cronon (1991) and Needham (2014).

40. As Ruth Morgan's analysis of rhetoric regarding water "independence" in San Diego and Perth, Australia, suggests, this sort of mismatch is not particularly unusual. See Morgan (2020). Calls for reducing dependence on faraway sources can often ring hollow.

41. See Ormerod (2019).

42. Ormerod (2019, 645).

43. Meilinger and Monstadt (2022a, 17).

44. See Hundley, Jackson, and Patterson (2016). Mulholland was at the time the chief engineer of LA's Bureau of Water Works and Supply, one of the two agencies that merged to become the LADWP in 1937.

2. DISRUPTING WATER CONSUMPTION AND CONTROL AT HOME

1. During my fieldwork period, the LADWP offered a rebate of $1.75 per square foot for turf removal. During 2014 and the first half of 2015, Metropolitan contributed an additional $2 per square foot for its member agencies, including LADWP. Though Metropolitan allocated $450 million for its program, the funds were exhausted in July 2015, cutting the rebate within the city of LA. In the years since, this figure has been increased several times, most recently to $5 per square foot for residential customers in 2022. The choice to offer these subsidies was

grounded in the recognition that irrigation represented a tremendous portion of single-family home water use in LA in the early 2010s—a full 54 percent, per one high-profile study (Mini, Hogue, and Pincetl 2014). The same agencies' hesitation to encourage domestic-scale wastewater reuse stands in contrast to this form of public investment in reducing outdoor water consumption.

2. See Baptista (2015), Anand (2020), and Chelcea (2023).

3. See Radonic (2019a, 2019b) and Vine (2018).

4. See Sofoulis (2005). See Allon and Sofoulis (2006), Karvoven (2011), Strang (2004, 2009), and Woelfle-Erskine (2015). Notably, a distinct but complementary body of critical scholarship questions the utility of focusing on centralized water networks and their attendant spatial logics in cities dominated by alternative provisioning arrangements. See Meehan (2014), Furlong and Kooy (2017), and Truelove (2019). While the context explored here approximates the "modern infrastructural ideal" of universal provision, reading the case alongside such work helps to further denaturalize many of the assumptions about the water grid that water managers articulate throughout this chapter.

5. See Sklar (2008) and Sharpsteen (2010). Having lived a few miles from Hyperion during the plant's notorious summer 2021 breakdown and associated sewage spill (see Lopez 2022), I can confirm that the plant is also occasionally a menace to local air quality.

6. Scaramelli (2019, 389).

7. See Maurer (2020) and Meilinger and Monstadt (2023).

8. See Singh (2013), Radonic (2019a), and Sony and Krishnan (2023).

9. See Baviskar (2020). As our instructors at the greywater installation course emphasized, the materials required to build a simple greywater system are not expensive. With guidance and persistence, a person with some home plumbing repair experience could install such a system. But for many—even those highly motivated by the prospect of domestic water reuse—the task is complicated enough to seem prohibitive. The few people I encountered during fieldwork who mentioned a DIY greywater system all told me that they had simply disconnected the hose of their laundry machine from the sewer and redirected it to their garden, foregoing the work of burying pipes or digging mulch basins. While not discounting the reality that some people (such as the Greywater Guerrillas in the 2000s, as described in this chapter) auto construct long-term, low-cost greywater reuse infrastructures, my characterization of the high cost and class privilege commonly associated with these infrastructures reflects most greywater adoptees I observed within LA during my fieldwork period.

10. Dicum (2007).

11. See Meehan (2014) and Ormerod (2019).

12. See Ingham (1980).

13. See City of Los Angeles Bureau of Water Reclamation (1992). Subsequent studies, undertaken by a mix of university researchers, water agencies,

and greywater advocacy groups, have reiterated these findings, indicating that long-term greywater reuse has no substantive negative impacts on soil, plant, or human health. See Allen, Bryan, and Woelfle-Erskine (2013), Lu et al. (2013), and Roesner et al. (2006). While the research suggests that long-term greywater irrigation is occasionally associated with increased soil saline levels (likely due to the salts in mainstream US detergents), virtually no other problems were observed on test sites.

14. Woelfle-Erskine et al. (2006). Co founder Laura Allen has also authored a pair of books—*The Water-Wise Home* (2015) and *Greywater Green Landscape* (2017)—that provide greywater installation instructions and further elaboration on the rationale behind its installation.

15. While the state standard is now relatively permissive, individual jurisdictions retain the capacity to enact stricter rules.

16. See Allen, Bryan, and Woelfle-Erskine (2013).

17. See Isenhour, McDonough, and Checker (2015).

18. See Isenhour (2011); see also Von Schnitzler (2008).

19. See Allon and Sofoulis (2006), Sofoulis (2005), and Strang (2009).

20. See Ormerod (2019).

21. My purpose here is not to overstate or romanticize the water management capacity of LA residents. Indeed, in one interview with a greywater adoptee who worked as a professional plumber, I was treated to a litany of stories about old, unsafe, poorly maintained greywater systems that the interlocutor had encountered in his professional life. Maintenance failures unquestionably occur. I suggest, however, there's some distance between acknowledging this reality and assuming its inevitably, the leap commonly made in water manager characterizations of these systems. Highlighting such pessimistic characterizations, I seek to surface the dynamic of assumed continuity and control that underpins them.

22. See Maurer (2020). This analysis of gardening practices in deindustrial Michigan draws helpful attention to the scalar dimensions of such arrangements in what she terms the "social reproduction of the urban environment" (717). Highlighting the contrasting forms of relationality emerging around gardening within private yards and community gardens, the account demonstrates the limitations of attempts to realize "just sustainability" through home-based ecological practices.

23. See Vine (2018).

24. See Vine (2018, 414).

25. See Checker (2011, 2020) and Newman (2011, 2015).

26. Perhaps unsurprisingly, given their general skepticism toward distributed infrastructures, several of my water manager interlocutors described such claims about the water table as dubious—particularly in areas outside the northeastern Valley. Nonetheless, I include the claim here to emphasize the sense of extended ecological benefits that Hank and the other greywater advocates sought to script for the technology.

27. See Williams (2019).

28. See Zhang (2020).

3. MANAGING LANDSCAPE

1. See LADWP (2021, 5–9).

2. Recent critical writing by geographers Joshua Cousins, Valentin Meilinger, and Jochen Monstadt explores key dimensions of LA's efforts to capture its stormwater. See Cousins (2017a, 2017b, 2017c) and Meilinger and Monstadt (2022b). While specific arguments and points from their publications are cited throughout this manuscript, I want to acknowledge that my analysis throughout the following chapters builds on their accounts of LA's ongoing work to rationalize its runoff.

3. While figures vary, one study my interlocutors cited frequently indicates that, through the early 1960s, more than 80 percent of the region's rainfall was absorbed into the ground or evaporated. Now, due to the impervious surfaces widespread urbanization have brought, this figure is closer to 50 percent. See Piechota and Dallman (2010).

4. This rich and growing body of work includes a range of case studies from urban and rural environments across the globe, demonstrating the growing footprint of such infrastructural nature projects. See Carse (2012, 2014), Wakefield and Braun (2019), Nelson et al. (2020), Wakefield (2020), Kurtiç (2023), and Lock (2023), for examples.

5. Carse (2012, 540).

6. See Roberts (2013), Koslov (2016), and Zeiderman (2016).

7. The growing, interdisciplinary body of literature on the pursuit of urban resilience has foregrounded experiences of and plans to mitigate localized flood risks. See Blok (2016), Colven (2017), Goh (2021), Wakefield (2020), and Ley (2021). While attending to the history of flooding in the northeastern Valley (which, as the narrative presented here attests, has shaped the landscape's development trajectory), I understand the project of managing this area for the purposes of water supply production as a related-but-distinct project from initiatives centered on flood risk. In many ways this approach is even more ambitious, assuming a human capacity to design distributed infrastructures capable of mitigating potential flood damage *while* capturing potential floodwaters for use as a water resource.

8. See Davis (1990), Gumprecht (2001), Desfor and Keil (2004), Orsi (2004), and Gandy (2006, 2014).

9. See Randle (2021).

10. Of these five sites, the LADWP owns only the Tujunga facilities. The northeastern Valley's other four spreading grounds—Pacoima, Branford, Hansen, and Lopez—were all developed by and are currently managed by the LA

County Flood Control District. Per the LADWP, the Tujunga facility is "operated by LACFCD in partnership with LADWP" (LADWP 2021, 6-2). As the only LADWP-owned facility, the Tujunga Spreading Grounds are sometimes used to spread and store "excess" LA Aqueduct water piped down from the Owens Valley, a function for which the other grounds are not operated. As I discuss elsewhere, while the LADWP has considered pursuing additional subterranean storage arrangements for LA Aqueduct water in sites located beyond city borders, there are no formal plans for such arrangements at this time. See Randle (2022a).

11. In this sense, the grounds present an interesting contrast with the "invisibility" of groundwater basins themselves as infrastructural natures, a condition that can guide infrastructural investments in a manner that prioritizes surface water projects like dams, due to the wider public recognition they tend to garner. See Colven (2020).

12. See LADWP (2021, 6-3).

13. See Redfield (2000) for an elaboration of the "shadow history" concept.

14. See Mitchelson (1930).

15. See *Los Angeles Times* (1899).

16. There is no dedicated flood management agency within the city of LA.

17. See Los Angeles County Flood Control Act (1915, 1502).

18. See *Los Angeles Times* (1931).

19. See Orsi (2004) and Cobery (2012).

20. Other areas of the Valley were annexed during the years that followed, also to procure access to aqueduct water. See Barraclough (2011).

21. See Hoffman (1981), Kahrl (1982), and Scavo (2013).

22. See Lane (1934) regarding the Tujunga Facility. The 1931 founding of the Metropolitan Water District facilitated a similar trajectory beyond the city's borders. That September, the same month that Flood Control adopted its comprehensive plan, Southern California voters approved a $220 million bond measure to fund the construction of Metropolitan's Colorado River Aqueduct. The pipeline's water first reached the Southland in 1941, marking a substantial shift in the region's water provision calculus. The City of LA no longer had a monopoly on faraway water. Now, any water jurisdiction that joined Metropolitan could supply its customers with imports. In 1952 Metropolitan adopted the "Laguna Declaration," a pledge to offer a permanent water supply for the region and to provide water to any agency that requested it within Metropolitan's three-county service area. See Zetland (2009). In support of this pledge, the agency contracted with the state government to provide water from Northern California through the State Water Project aqueducts, which connected to Metropolitan's system in 1973. As a result, water remained plentiful and cheap across the region during its post–World War II boom years, further diminishing the sense of urgency about the project of managing stormwater as a resource.

23. See Scott (2006) and Molle (2009).

24. See O'Neill (2006).

25. *Los Angeles Times* (1968).

26. *Los Angeles Times* (1968).

27. See LADWP (2016, 2021.

28. See City of LA v. City of San Fernando, et al., 14 Cal. 3d 199, 123 Cal. Rep. 1 (1975). While reaffirming LA's pueblo rights, the California Supreme Court acknowledged that a lower court's assertion that the claim had no legitimate basis in law was not necessarily wrong—but argued that the right should nonetheless be upheld based on *stare decisis*. See Hundley (2001, 335). Which is to say: LA got lucky on this one.

29. As legal scholars have noted, it also affirms LA's exclusive right to Owens Valley water that enters the aquifer via the spreading grounds. See Gleason (1976, 1977) and Taguchi (2003). However, as LA enjoys claims on other sites for storing that water (and continues to explore additional ones) and has no other large-scale option for its stormwater storage, "filling the storage tank" with stormwater is widely seen as the most advantageous option for the city.

30. In addition to its adjudicated right to pump up to 87,000 acre-feet from the San Fernando Basin each year, the City of LA also has the right to pump up to 4,070 acre-feet combined from the nearby Sylmar and Eagle Rock Basins. Currently the city has two wells in operation within the Sylmar Basin and no production in the Eagle Rock Basin.

31. Not coincidentally, such lands are also often marked by underdeveloped infrastructure and disproportionate concentrations of environmental hazards. For excellent analyses of such dynamics across the US West, see Piper (2006), Needham (2014), Voyles (2015), and Powell (2018).

32. See Barraclough (2011).

33. See Barraclough (2011, 29).

34. Over the years many writers (including, unfortunately, this one; see Randle 2020) have reproduced the erroneous notion that the name Pacoima was adapted from a Tongva or Tataviam word for "rushing waters." This is a misconception. See Salinas (2019),

35. See Roderick (2001).

36. See Sides (2003, 104).

37. See de Guzman (2014a, 2014b). Japanese American farmers leased or owned substantial swathes of the northeastern Valley, concentrated around the neighborhoods of Sunland and Tujunga, throughout the 1920s and 1930s. While the community's agricultural presence was permanently diminished within the Valley due to land appropriation following the forced internment of the community during the US participation in World War II, many returned following the conflict to settle around Pacoima and the nearby city of San Fernando. See Barraclough (2011, 130–31).

38. See Sides (2003).

39. See Nadeau (1960) and Roderick (2001).

40. See Sklar (2008, 113).

41. Ponce ([1992] 2006, 3–5).

42. See Lubas (1978).

43. See Braxton (1986).

44. See Barraclough (2008).

45. See Scott (1996).

46. See Pulido (2000) and Barraclough (2011).

47. See Lejano and Ericson (2005), Maida (2011), and Shamasundar et al. (2015).

48. See Green (2008).

49. See Levin (1985b) and Ryan (1985).

50. See Stein (1983).

51. See Stein (1983).

52. See Levin (1985a).

53. As of November 2024, the LADWP had not announced the completion of the remediation. Notably, in October 2024 the EPA announced a settlement with Honeywell International Inc. to fund the North Hollywood remediation facilities. Operations at a manufacturing facility owned by a Honeywell predecessor had caused substantive groundwater contamination, according to regulators. The company's financial contribution to the remediation facilities' construction was not publicized at the time of the announcement. See James (2024).

54. See Powis (2021).

55. See LADWP (2016, 62).

56. See LADWP (2021, 5–8).

57. See Green (2006).

58. US EPA (2019).

59. Key antecedents include the holistic urban watershed planning approaches formulated in response to the ecological effects of post–World War II suburbanization. As Adam Rome details in his history of US environmentalism, common mid-twentieth-century building practices such as bulldozing lots before construction compacted soils in a manner that made the lawns planted on them behave "more like slabs of concrete than sponges" (2001, 197). As the hydrological effects of such development quickly accumulated across the nation, in the form of increased flooding and polluted watercourses, the US Geological Survey began to study the topic during the 1960s. This initiative led to publication of Luna Leopold's *Hydrology for Urban Land Planning* (1968), which laid out principles for planning new developments in a manner that maximized the landscape's stormwater capture capacity. At the end of the decade, these issues leapt from the circles of developers and engineers into the mainstream, with the publication of Ian McHarg's *Design with Nature* (1969). Influenced by the work of nineteenth-century landscape designer Frederick Law Olmsted and

early twentieth-century urban planner Ebenezer Howard, the text argues for the importance of maximizing local ecological functions for human-determined goals, a notion that gained traction quickly in landscape architecture programs. See Karvoven (2011).

60. Notably, at the public meeting about this project at which I met Kumal, most of the attendees' comments were related to maintenance concerns, as the owners of the properties abutting those parkway strips were expected to take care of the basins' long-term upkeep.

61. During my months of LA-based fieldwork, several interactive stakeholder workshops focused on water were advertised as *charrettes*, defined as "a meeting or public workshop devoted to a concerted effort to solve a problem or plan the design of something" in the *Oxford English Dictionary*. Such meetings tended to feature at least one small group exercise that involved drawing on a map of some section of the city to identify opportunities to retrofit new infrastructures for stormwater capture.

62. See Cronon (1991), Harvey (1996), Gandy (2002b), Brenner and Schmid (2013), and Arboleda (2016).

4. INFRASTRUCTURAL WORK AND INFILTRATING RUNOFF

1. Anand (2011, 543).

2. See Barry (2013), Gidwani (2015), Fredericks (2018), and Stokes and De Coss-Corzo (2023).

3. See Elyachar (2010) on phatic labor, the ongoing social work of connection that frequently buttresses infrastructural arrangements.

4. See Callon ([1983] 1999) and Latour (1987).

5. See Gupta (2018) and Stamatopoulou-Robbins (2021). See also Choy and Zee (2015).

6. In contrast with anthropological accounts of "connectionwork" that emphasize a personal, embodied sense of relation with other beings or an imagined community (see Appel 2021, 2023), here I focus on the recognition of material links within an ecological system.

7. My approach to this concept is particularly influenced by Andrea Ballestero's (2019b) writing on groundwater materialities. See also Weizman (2002) and Elden (2013)

8. See Vine (2018) and Ojani (2023).

9. See Jensen and Morita (2015), Shapiro-Garza et al. (2020), and Suarez (2023).

10. Dempsey and Robertson (2012, 768).

11. See McElwee (2017), Nost (2022), and Nost and Goldstein (2022).

12. See Robertson (2006).

13. Robertson (2006, 377).

14. Drawing on stakeholder interviews conducted around the same time as my own, Joshua Cousins documents a similar pattern within LA, including the perspective of one LADWP worker who bluntly stated: "I strongly agree that centralized projects are much better at handling stormwater cost-effectively compared to distributed projects and LID [low-impact development]. This is easily seen as a cost-benefit analysis when considering [the] amount of water capture [for flood control] and infiltrated [for water supply]" (Cousins 2017a, 40).

15. See Finewood (2016, 1001).

16. See Finewood (2016) and Finewood, Matsler, and Zivkovich (2019).

17. See Cousins (2017b, 2017c) and Meilinger and Monstadt (2022b).

18. See Kroepsch and Clifford (2022). See also Ballestero (2019b), Bessire (2022), and Walsh (2022).

19. See Turnhout et al. (2014, 582) cited in McElwee (2017). See also Sullivan (2018).

20. Following anthropological precedent, my use of the word *value* is largely aligned with the definition "how much someone is willing to give up to obtain something"—as opposed to the term *values*, understood to connote "moral understandings of what is right and good" (du Bray et al. 2019, 21); see also Kluckhohn (1961), Appadurai (1988), and Graeber 2001). As the section's analysis attests, however, in this context the contours of the former category are ambiguous and actively contested, and sometimes bleed into the latter.

21. See Robertson (2007, 2012), Collard and Dempsey (2013), and Bridge et al. (2020). See also Gómez-Baggethun and Ruiz-Pérez (2011) for an economic framing.

22. See Huff (2021). See also Cavanaugh and Benjaminsen (2017) and Osborne and Shapiro-Garza (2018).

23. See Bakker (2003, 2005).

24. See Matsler (2019), Cousins and Hill (2021), and Tozer et al. (2023). As discussed in chapter 5, negotiations over how to assign value to environmental services also mark rural payments for ecosystem services (PES) schemes. But those configurations of infrastructural nature tend to be approached and managed as spatially extensive land conservation projects and are thus rarely governed by the same grey institutions and paradigms. In many such arrangements, a funder—consisting of some combination of public sector and market actors—will provide financing intended to incite rural residents' conservation practices within a given landscape. As such, the valuation of local conservation labor, rather than infrastructure construction, tends to be the key point of engagement. Approached as water supply infrastructure to be developed by a public water utility, LA's distributed infiltration installations thus seek to realize nature's labor in a context wherein the land's provisioning work (i.e., water

supply augmentation) is more directly imbricated with established calculative practices.

25. In this sense, we might say that they hoped to secure what anthropologist Emily Brooks terms a "number narrative," a story "about how an environmental object works in a particular place and time" that serves to hold "waterworlds" (Hastrup 2014) of people, places, things, and practices together (Brooks 2017, 34).

26. Notably, in contrast to the SCMP, the US Bureau of Reclamation's (2016) "Trade-off Analysis & Opportunities" section of its *Los Angeles Basin Study: The Future of Stormwater Conservation*—discussed at some length later in this chapter—includes an extended section on reliability analysis. Also, for a fuller discussion of water reliability and temporality in LA, see chapter 1.

27. See Jackson and Palmer (2015).

28. See McElwee (2017). Pamela McElwee's detailed review of ecosystem services metrologies, published two years after I conducted this (far less systematic) search, reflects a similar state of the field and is an excellent resource for readers interested in the topic.

29. In this case, the efforts bore fruit, as the final report, published the following year, included ecosystem services valuations and a substantive discussion on the best way to derive them. See US Bureau of Reclamation (2016, 23–26, 39–46).

30. The piecemeal nature of stormwater project funding in LA makes the total figure of local stormwater investment difficult to ascertain. While the SCMP offers a detailed accounting of recent and proposed stormwater projects in the city, it outlines funding principles—prioritize projects rated as cost-efficient, pursue grants and partnerships with other public agencies—but gives no detailed accounting of total investments. The closest the report comes to estimating future investment from the water department comes in a discussion of modeled future scenarios, which assumed annual expenditures between $15 million and $35 million and project cos tsharing with other entities ranging from 30 to 70 percent. See LADWP (2015, 61–62). Such ambiguity regarding investments is replicated in the LADWP's 2020 UWMP, which lists estimated costs for some in-process and planned projects but provides no such accounting of completed projects. See LADWP (2021, 6–13).

31. See Robbins (2007), Chowdhury et al. (2011), Hondagneu-Sotelo (2014), and Meilinger and Monstadt (2023).

32. As this description suggests, these infrastructures embody an interesting intersection of what David Wachsmuth and Hilary Angelo term the ideologies of "green" and "gray" urban natures: "Green urban nature is the return of nature to the city in its most verdant form, whereas gray urban nature is the concept of social, technological urban space as already inherently sustainable" (2018, 1039). In the form of these distributed infrastructures, nature takes on resonances associated with both past and future.

33. See Cousins (2017a, 2017b, 2017c) and Meilinger and Monstadt (2022a).

34. Finewood (2016, 1001–2).

35. Despite Keith's assertions, there is no broadly accepted legal require-ment for this conservative approach. In 1996 California voters passed Propo-sition 218, a ballot measure that severely circumscribes water price structures across the state by prohibiting public agencies from charging more for ser-vices than their actual cost. Under its aegis, water agency attempts to develop schemes that charge households extremely high rates for water consumed be-yond a certain volume have been deemed illegal. But there is no such clear-cut law prohibiting agencies from investing in expensive water sources. As a staffer in LA's Office of the Ratepayer Advocate once explained it to me: "The DWP Board of Commissioners in their infinite wisdom could approve any number of stormwater projects with huge inefficiencies because the mayor told them to. The money would have to come from the water rates. . . . Only people like this office, [Keith], and the charter prevent that from happening. There's no legal impediment."

36. Many hope that the 2018 passage of LA County's Measure W, a parcel tax on stormwater intended to fund runoff-capture infrastructures across the coun-ty's eighty-five cities to the tune of $280 million annually, will eventually ease the process of pursuing such funding. While providing a substantial new source of public money for these projects, by its nature the arrangement does not ad-dress the divergent institutional logics and priorities discussed here.

5. ECOSYSTEM DUTIES AND ENVIRONMENTAL JUSTICE

1. Over the years I've found that comparing domestic stormwater recharge installations with rooftop solar panels helps to clarify a key element of these ar-rangements. Solar panels, another distributed infrastructure that produces new resources within the space of the city, can directly provide a household with elec-tricity, substantially lowering its utility bills. In contrast, the recharge infra-structures discussed here bolster only the groundwater supply that serves the whole city. This scalar mismatch between the work and benefits associated with these infrastructures is constitutive of the dynamics explored throughout this chapter.

2. Danielle DiNovelli-Lang and Karen Hébert (2018) helpfully define the term *ecological labor* as signaling both the bodily exertions and the affective, per-formative aspects that such ecologically reproductive work entails—an under-standing that has shaped this book's approach to my ethnographic material. Regarding another key piece of terminology, in the anthropological literature, the phrase "environmental justice" is far more common than "environmental

equity" or "environmental equality," a choice that echoes that of activists who, from the 1980s onward, saw "environmental justice" as the more inclusive, appropriately politicizing phrase to characterize a wide range of socio-environmental harms. Following Laura Pulido (2000), here I use the terms *equity* and *inequity* with reference to allocation issues, which I understand as a key dimension of broader environmental justice struggles and theorizations.

3. To reiterate, Scaramelli uses the phrase "moral ecology" to signal "people's notions of just relations between people, land, water, and nonhuman animals, plants, buildings, technologies, and infrastructures" (2019, 389).

4. See McElwee (2016), Greenleaf (2020a, 2020b), and Nelson, Bremer, and Meza Prado (2020).

5. See Joslin (2020), Hoag (2022), and Kurtiç (2023).

6. See Greenleaf (2021).

7. See Everett et al. (2015), Qiao et al. (2019), and Zuniga-Teran et al. (2021).

8. See Robertson (2012).

9. See Federici (1975) and Molyneux (1979).

10. See Moore (2015).

11. See di Leonardo (1998), Rosaldo and Lamphere (1974), and Cielo, Coba, and Vallejo (2016).

12. See Rosenbaum (2017).

13. See Collins (2016) and Muehlebach (2012).

14. See Wynter (1992).

15. See Hawthorne (2019)and McKittrick (2013).

16. See DiNovelli-Lang and Hébert (2018).

17. See Parreñas (2018), Moore (2019), and Brondo (2021).

18. See Bullard (1990), Holifield (2001), Mohai, Pellow, and Roberts (2009), Taylor (2014), Pulido (2016), and Kim (2021).

19. See Barraclough (2009), Park and Pellow (2011), Sister, Wolch, and Wilson (2010), and Newman (2015).

20. See Checker (2011, 2020), Dooling (2009), Wolch, Byrne, and Newell (2014), and Anguelovski, et al. (2018).

21. See Anand, Gupta, and Appel (2018), Barnes (2014), De Coss-Corzo (2020), Ernwein (2020), and Günel (2019) for particularly compelling ethnographic accounts of such essential maintenance.

22. See Anand (2012) and Von Schnitzler (2016).

23. See Barnes (2017).

24. See Larkin (2008), Collier (2011), Kinder (2016), and Alexandre (2018).

25. See Park, Stenstrom, and Pincetl (2009).

26. Owens, Queally, and Reyes (2014).

27. As with the urban infrastructural nature projects analyzed here, scholars studying rural PES schemes have demonstrated that such arrangements rework relations between rural residents and the state, tightening such links even in

the absence of formal land acquisition practices. See Bleeker and Vos (2019) and Hommes, Boelens, et al. (2019), and Hommes, Hoogesteger, et al. (2022).

28. See Meilinger and Monstadt (2023).

29. While several NGOs were working on stormwater recharge projects in the northeastern Valley during my research period, few of their staff members lived in these neighborhoods. In general, the employees of these organizations were whiter, wealthier, and more educated than the project participants. These gaps between the groups crafting and the groups targeted for these interventions marked the dynamics I observed both in the neighborhood and when NGO workers discussed the projects in policy settings.

30. Such framings clearly reflect a turn toward what Erik Swyngedouw terms "governance-beyond-the-state" characteristic of many neoliberal policy "innovations"—albeit in far more celebratory terms than a critical geographer would use to describe them. See Swyngedouw (2005).

31. Readers familiar with Gökçe Günel's *Spaceship in the Desert* (2019) may notice resonances to the "man with a brush," a phrase meant to evoke poorly paid immigrant workers tasked with wiping off the dust that impairs the functioning of solar panels in Masdar City, a desert eco-city under construction in the United Arab Emirates. Though acknowledged in some contexts as essential to the project's functioning, Günel observes that "man with a brush" embodied the sort of cheap, low-tech human labor the city's promoters intentionally excluded from the techno-utopian public narrative favored by the city's promoters (2019, 59–61). As in the LA context, such framings don't linger on the messy, quotidian work of infrastructure maintenance.

32. See DiNovelli-Lang and Hébert (2018), Moore (2019), Brondo (2021), Mills-Novoa (2023), and Muehlebach (2012).

33. See Gandy (2002a).

34. See Kinder (2016) and Gaber (2021).

35. While salaried work at self-described environmental justice organizations has grown more common in recent years (and environmental justice work portfolios have grown more common within public agencies), it bears noting that articulations of environmental justice priorities within a given landscape can vary wildly. See Brodkin (2009), Harrison (2019), and Kim (2021).

36. See Parreñas (2018).

37. See Checker (2005), Pulido (2000), Kim (2021), and Taylor (2014).

38. Of course this literature does not suggest that such ecological labor-based redistribution arrangements are always smooth or adequate to the needs of residents—or even necessarily successful in cultivating the desired ecosystem services. See Greenleaf (2020a) and Hoag (2022). With this in mind, I highlight such programs to suggest a productive starting point for discussions of infrastructural natures: one that foregrounds their potential to support livelihoods.

EPILOGUE

1. Smith (2023).

2. See Sahagun (2023).

3. See Crow-Miller, Webber, and Molle (2017).

4. See Vartabedian (2022), Los Angeles Waterkeeper (2023), and Smith (2023).

5. That interim sounds remarkably short when compared to the current estimate for the city's much-touted Hyperion recycling project, which the LADWP currently predicts will be completed around 2058.

6. See White (1996a).

7. See White (1996a, 1996b).

8. See Dove (1983) and Fairhead and Leach (1996).

References

Addie, Jean-Paul D. 2022. "The Times of Splintering Urbanism." *Journal of Urban Technology* 29 (1): 109–16.

Agrawal, Arun. 2005. *Environmentality: Technologies of Government and the Making of Subjects*. Durham, NC: Duke University Press.

Alexandre, Kessie. 2018. "When It Rains: Stormwater Management, Redevelopment, and Chronologies of Infrastructure." *Geoforum* 97 (December): 66–72.

Allen, Laura. 2015. *The Water-Wise Home: How to Conserve, Capture, and Reuse Water in Your Home and Landscape*. North Adams, MA: Storey Publishing.

———. 2017. *Greywater, Green Landscape: How to Install Simple Water-Saving Irrigation Systems in Your Yard*. North Adams, MA: Storey Publishing.

Allen, Laura, Sherry Bryan, and Cleo Woelfle-Erskine. 2013. *Residential Greywater Reuse Systems in California: An Evaluation of Soil, Water Quality, User Satisfaction, and Installation Costs*. Berkeley, CA: Greywater Action in collaboration with City of Santa Rosa and Ecology Action of Santa Cruz.

Allon, Fiona, and Zoë Sofoulis. 2006. "Everyday Water: Cultures in Transition." *Australian Geographer* 37 (1): 45–55.

Amironesei, Razvan, and Caleb Scoville. 2019. "Groundwater in California: From Juridical and Biopolitical Governmentality to a Political Physics of Vital Processes." *Theory, Culture and Society* 36 (5): 133–57.

Anand, Nikhil. 2011. "Pressure: The Politechnics of Water Supply in Mumbai." *Cultural Anthropology* 26 (4): 542–64.

———. 2012. "Municipal Disconnect: On Abject Water and Urban Infrastructure." *Ethnography* 13(4): 487–509.

———. 2017. *Hydraulic City: Water and the Infrastructures of Citizenship in Mumbai.* Durham, NC: Duke University Press.

———. 2020. "Consuming Citizenship: Prepaid Meters and the Politics of Technology in Mumbai." *City and Society* 32 (1): 47–70.

Anand, Nikhil, Akhil Gupta, and Hannah Appel, eds. 2018. *The Promise of Infrastructure.* Durham, NC: Duke University Press.

Anguelovski, Isabelle, James Connolly, Laia Masip, and Hamill Pearsall. 2018. "Assessing Green Gentrification in Historically Disenfranchised Neighborhoods: A Longitudinal and Spatial Analysis of Barcelona." *Urban Geography* 39(3): 458–91.

Appadurai, Arjun, ed. 1988. *The Social Life of Things: Commodities in Cultural Perspective.* Cambridge: Cambridge University Press.

Appel, Ariel. 2021. "Connectionwork: 'Connection' in the Practices and Perspectives of Israeli Neoforagers." *HAU: Journal of Ethnographic Theory* 11 (2): 551–66.

———. 2023. "Unruly Spaces, Unsettling Transformations: Nature Connection, Neoforaging, and Unmediated Encounters with Others in Israel/Palestine." *Environment and Planning E: Nature and Space* (March): 25148486231160080.

Arboleda, Martín. 2016. "In the Nature of the Non-City: Expanded Infrastructural Networks and the Political Ecology of Planetary Urbanisation." *Antipode* 48 (2): 233–51.

Bacigalupi, Paolo. 2015. *The Water Knife.* New York: Knopf.

Bakker, Karen J. 2003. *An Uncooperative Commodity: Privatizing Water in England and Wales.* Oxford: Oxford University Press.

———. 2005. "Neoliberalizing Nature? Market Environmentalism in Water Supply in England and Wales." *Annals of the Association of American Geographers* 95 (3): 542–65.

Ballestero, Andrea. 2019a. *A Future History of Water.* Durham, NC: Duke University Press.

———. 2019b. "The Underground as Infrastructure? Water, Figure/Ground Reversals, and Dissolution in Sardinal." In *Infrastructure, Environment, and Life in the Anthropocene*, edited by Kregg Hetherington, 17–44. Durham, NC: Duke University Press.

———. 2019c. "Touching with Light, or, How Texture Recasts the Sensing of Underground Water." *Science, Technology, & Human Values* 44 (5): 762–85.

Baptista, Idalina. 2015. "'We Live on Estimates': Everyday Practices of Prepaid Electricity and the Urban Condition in Maputo, Mozambique." *International Journal of Urban and Regional Research* 39 (5): 1004–19.

Barnes, Jessica. 2014. *Cultivating the Nile: The Everyday Politics of Water in Egypt.* Durham, NC: Duke University Press.

——. 2016. "Uncertainty in the Signal: Modelling Egypt's Water Futures." *Journal of the Royal Anthropological Institute* 22 (S1): 46–66.

——. 2017. "States of Maintenance: Power, politics, and Egypt's Irrigation Infrastructure." *Environment and Planning D: Society & Space* 35 (1): 146–64.

Barraclough, Laura R. 2008. "Rural Urbanism: Producing Western Heritage and the Racial Geography of Postwar Los Angeles." *Western Historical Quarterly* 39 (2): 177–202.

——. 2009. "South Central Farmers and Shadow Hills Homeowners: Land Use Policy and Relational Racialization in Los Angeles." *Professional Geographer* 61 (2): 164–86.

——. 2011. *Making the San Fernando Valley: Rural Landscapes, Urban Development, and White Privilege.* Athens: University of Georgia Press.

Barry, Andrew. 2013. *Material Politics: Disputes Along the Pipeline.* New York: Wiley.

Barua, Maan. 2023. *Lively Cities: Reconfiguring Urban Ecology.* Minneapolis: University of Minnesota Press.

Battistoni, Alyssa. 2017. "Bringing in the Work of Nature." *Political Theory* 45 (1): 5–31.

Baviskar, Amita. 2020. *Uncivil City: Equity and the Commons in Delhi.* New Delhi: Sage Publications.

Bennett, Jane. 2010. *Vibrant Matter: A Political Ecology of Things.* Durham, NC: Duke University Press.

Bertenthal, Alyse. 2021. "The Alchemy of the Public Interest." *Yale Journal of Law & the Humanities* 32 (1): 1–37.

Besky, Sarah, and Alex Blanchette, eds. 2019. *How Nature Works: Rethinking Labor on a Troubled Planet.* Santa Fe, NM: School for Advanced Research Press.

Bessire, Lucas. 2022. "Aquifer Aporias Toward a Comparative Anthropology of Groundwater Depletion." *Current Anthropology* 63(3): 350–59.

Björkman, Lisa. 2015. *Pipe Politics, Contested Waters: Embedded Infrastructures of Millennial Mumbai.* Durham, NC: Duke University Press.

——. 2018. " The Engineer and the Plumber: Mediating Mumbai's Conflicting Infrastructural Imaginaries." *International Journal of Urban and Regional Research* 42(2): 276–94.

Bleeker, Sonja, and Jeroen Vos. 2019. "Payment for Ecosystem Services in Lima's Watersheds: Power and Imaginaries in an Urban-Rural Hydrosocial Territory." *Water International* 44 (2): 224–42.

Blok, Anders. 2016. "Assembling Urban Riskscapes: Climate Adaptation, Scales of Change and the Politics of Expertise in Surat, India." *City* 20 (4): 602–18.

Blomquist, William A. 1992. *Dividing the Waters: Governing Groundwater in Southern California.* San Francisco: ICS Press.

Borgias, Sophia. 2024a. "Denaturalizing Dispossession in the Political Ecology of the American West: Reassessing the History of the Los Angeles Aqueduct and Its Implications for Indigenous Land and Water Rights." *Annals of the American Association of Geographers* 114 (6): 1232–50.

———. 2024b. "Drought, Settler Law, and the Los Angeles Aqueduct: The Shifting Political Ecology of Water Scarcity in California's Eastern Sierra." *Journal of Political Ecology* 31 (1).

Boyer, Dominic. 2024. "Infrastructural Citizenship and Geosolidarity: Making Green Infrastructure in Petroliberal Houston." *American Ethnologist* 51(3): 338–49.

Braxton, Greg. 1986. "A Dump or Paradise? With 50 Junkyards, Sun Valley May Be the 'Auto Wrecking Capital' of California, but Many Residents Brag of a Small-Town Ambiance." *Los Angeles Times*, May 19, 1986.

Brenner, Neil, and Carl Schmid. 2013. "The 'Urban Age' in Question." *International Journal of Urban and Regional Research* 38 (3): 731–55.

Bridge, Gavin, Harriet Bulkeley, Paul Langley, and Bregje Van Veelen. 2020. "Pluralizing and Problematizing Carbon Finance." *Progress in Human Geography* 44 (4): 724–42.

Broadous, Hillary. 1977. Interview by William Huling, December 22. Department of History and University Library's Urban Archives Center, California State Northridge University.

Brodkin, Karen. 2009. *Power Politics: Environmental Activism in South Los Angeles.* New Brunswick, NJ: Rutgers University Press.

Brondo, Keri Vacanti. 2021. *Voluntourism and Multispecies Collaboration: Life, Death, and Conservation in the Mesoamerican Barrier Reef.* Tucson: University of Arizona Press.

Brooks, Emily. 2017. "Number Narratives: Abundance, Scarcity, and Sustainability in a California Water World." *Science as Culture* 26 (1): 32–55.

Bullard, Robert D. 1990. *Dumping in Dixie: Race, Class, and Environmental Quality.* Boulder, CO: Westview Press.

Butler, Octavia E. 1993. *Parable of the Sower.* New York: Four Walls Eight Windows Press.

Butt, Waqas. 2023. *Life Beyond Waste: Work and Infrastructure in Urban Pakistan.* Stanford, CA: Stanford University Press.

California Department of Water Resources. 2021. *Urban Water Management Plan Guidebook 2020.*

Callon, Michel. (1983) 1999. "Some Elements of a Sociology of Translation: Domestication of the Scallops and the Fisherman of St. Brieuc Bay." In *The Science Studies Reader,* edited by Mario Biagioli, 67–83. New York: Routledge.

Cantor, Alida. 2021. "Hydrosocial Hinterlands: An Urban Political Ecology of Southern California's Hydrosocial Territory." *Environment and Planning E: Nature and Space* 4 (2): 451–74.

Carse, Ashley. 2012. "Nature as Infrastructure: Making and Managing the Panama Canal Watershed." *Social Studies of Science* 42 (4): 539–63.

———. 2014. *Beyond the Big Ditch: Politics, Ecology, and Infrastructure at the Panama Canal.* Cambridge, MA: MIT Press.

Cavanagh, Connor Joseph, and Tor Arve Benjaminsen. 2017. "Political Ecology, Variegated Green Economies, and the Foreclosure of Alternative Sustainabilities." *Journal of Political Ecology* 24 (1): 200–216.

Cerra, Joshua. 2017. "Emerging Strategies for Voluntary Urban Ecological Stewardship on Private Property." *Landscape Urban Planning* 157: 586–97.

Chahim, Dean. 2022. "Governing Beyond Capacity: Engineering, Banality, and the Calibration of Disaster in Mexico City." *American Ethnologist* 49 (1): 20–34.

Chalfin, Brenda. 2023. *Waste Works: Vital Politics in Urban Ghana.* Durham, NC: Duke University Press.

Checker, Melissa. 2005. *Polluted Promises: Environmental Racism and the Search for Justice in a Small Southern Town.* New York: New York University Press.

———. 2011. "Wiped out by the 'Greenwave': Environmental Gentrification and the Paradoxical Politics of Urban Sustainability." *City and Society* 23 (2): 210–29.

———. 2020. *The Sustainability Myth: Environmental Gentrification and the Politics of Justice.* New York: New York University Press.

Chelcea, Liviu. 2023. "Catch-all Technopolitics." *American Ethnologist* 50: 260–73.

Chowdhury, Rinku Roy, Keli Larson, Morgan Grove, Colin Polsky, Elizabeth Cook, Jeffrey Onsted, and Laura Ogden. 2011. "A Multi-Scalar Approach to Theorizing Socio-Ecological Dynamics of Urban Residential Landscapes." *Cities and the Environment* 4 (1): 1–21.

Choy, Timothy, and Jerry Zee. 2015. "Condition—Suspension." *Cultural Anthropology* 30 (2): 210–23.

Christen, Kris. 2005. "Water Reuse: Getting Past the 'Yuck Factor.'" *Water Environment & Technology* 17 (11): 11–15.

Christophers, Brett. 2018. "Risk Capital: Urban Political Ecology and Entanglements of Financial and Environmental Risk in Washington, D.C." *Environment and Planning E: Nature and Space* 1 (1–2): 144–64.

Cielo, Cristina, Lisset Coba, and Ivette Vallejo. 2016. "Women, Nature, and Development in Sites of Ecuador's Petroleum Circuit." *Economic Anthropology* 3 (1): 119–32.

City of Los Angeles. 2015. *Sustainable City pLAn.* Los Angeles: City of Los Angeles.

———. 2018. *One Water 2040 Plan.* Los Angeles: City of Los Angeles.

———. 2019. *LA's Green New Deal: Sustainable City pLAn.* Los Angeles: City of Los Angeles.

City of Los Angeles Bureau of Water Reclamation. 1992. *Gray Water Pilot Project Final Report*. Los Angeles: City of Los Angeles.

Cobery, Art. 2012. *The Great Crescenta Valley Flood: New Year's Day 1934*. Mount Pleasant, SC: Arcadia Publishing Company.

Collard, Rosemary-Claire, and Jessica Dempsey. 2013. "Life for Sale? The Politics of Lively Commodities." *Environment and Planning A* 45 (1): 2682–99.

Collier, Stephen J. 2011. *Post-Soviet Social: Neoliberalism, Social Modernity, Biopolitics*. Princeton, NJ: Princeton University Press.

Collier, Stephen, and Andrew Lakoff. 2015. "Vital Systems Security: Reflexive Biopolitics and the Government of Emergency." *Culture & Society* 32 (2): 19–51.

———. 2021. *The Government of Emergency: Vital Systems, Expertise, and the Politics of Security*. Princeton, NJ: Princeton University Press.

Collins, Jane. 2016. "Expanding the Labor Theory of Value." *Dialectical Anthropology* 40:103–23.

Colven, Emma. 2017. "Understanding the Allure of Big Infrastructure: Jakarta's Great Garuda Sea Wall Project." *Water Alternatives* 10 (2): 250–64.

———. 2020. "Subterranean Infrastructures in a Sinking City: The Politics of Visibility in Jakarta." *Critical Asian Studies* 52(3): 311–31.

Cons, Jason. 2018. "Staging Climate Security: Resilience and Heterodystopia in the Bangladesh Borderlands." *Cultural Anthropology* 33 (2): 266–94.

Costanza, Robert, Ralph d'Arge, Rudolf de Groot, Stephen Farber, Monica Grasso, Bruce Hannon, Karin Limburg, et al. 1997. "The Value of the World's Ecosystem Services and Natural Capital." *Nature* 387 (6630): 253–60.

Cousins, Joshua J. 2017a. "Of Floods and Droughts: The Uneven Politics of Stormwater in Los Angeles." *Political Geography* 60:34–46.

———. 2017b. "Structuring Hydrosocial Relations in Urban Water Governance." *Annals of the Association of American Geographers* 107 (5): 1144–61.

———. 2017c. "Volume Control: Stormwater and the Politics of Urban Metabolism." *Geoforum* 85:368–80.

———. 2020. "Malleable Infrastructures: Crisis and the Engineering of Political Ecologies in Southern California." *Environment and Planning E: Nature and Space* 3 (3): 927–49.

Cousins, Joshua J., and Dustin T. Hill. 2021. "Green Infrastructure, Stormwater, and the Financialization of Municipal Environmental Governance." *Journal of Environmental Policy & Planning* 23 (5): 581–98.

Cronon, William. 1991. *Nature's Metropolis: Chicago and the Great West*. New York: W. W. Norton.

———. 1996. "The Trouble with Wilderness: Or, Getting Back to the Wrong Nature." *Environmental History* 1 (1): 7–28.

Crow-Miller, Britt, Michael Webber, and Francois Molle. 2017. "The (Re)turn to Infrastructure for Water Management?" *Water Alternatives* 10 (2): 195–207.

Daggett, Cara. 2019. *The Birth of Energy: Fossil Fuels, Thermodynamics, and the Politics of Work*. Durham, NC: Duke University Press.

Davis, Mike. 1990. *City of Quartz: Excavating the Future in Los Angeles*. New York: Verso.

——. 1998. *Ecology of Fear: Los Angeles and the Imagination of Disaster*. New York and London: Picador.

De Coss-Corzo, Alejandro. 2020. "Patchwork: Repair Labor and the Logic of Infrastructure Adaptation in Mexico City." *Environment and Planning D: Society and Space* 39 (2): 237–53.

de Guzman, Jean-Paul. 2014a. "And Make the San Fernando Valley My Home:" Contested Spaces, Identities, and Activism on the Edge of Los Angeles" (PhD diss., University of California, Los Angeles).

——. 2014b. "Race, Community, and Activism in Greater Los Angeles: Japanese Americans, African Americans, and the Contested Spaces of Southern California." In *The Nation and Its Peoples: Citizens, Denizens, Migrants*, edited by John S. W. Park and Shannon Gleeson, 29–48. New York: Routledge.

de La Bellacasa, Maria Puig. 2017. *Matters of Care: Speculative Ethics in More Than Human Worlds*. Minneapolis: University of Minnesota Press.

Dempsey, Jessica, and Morgan Robertson. 2012. "Ecosystem Services: Tensions, Impurities, and Points of Engagement within Neoliberalism." *Progress in Human Geography* 36 (6): 758–79.

Desfor, Gene, and Roger Keil. 2004. *Nature and the City: Making Environmental Policy in Toronto and Los Angeles*. Society, Environment, and Place. Tucson: University of Arizona Press.

Deverell, William, and Tom Sitton. 2013. "Forget It, Jake: Searching for the Truth in Chinatown." *Boom: A Journal of California* 3 (3): 3–7.

di Leonardo, Micaela. 1998. *Exotics at Home: Anthropologies, Others, American Modernity*. Chicago: University of Chicago Press.

Dicum, Gregory. 2007. "The Dirty Water Underground." *New York Times*, May 31, 2007, F1.

Diffenbaugh, Noah S., Daniel L. Swain, and Danielle Touma. 2015. "Anthropogenic Warming Has Increased Drought Risk in California." *Proceedings of the National Academy of Sciences* 112 (13): 3931–36.

DiNovelli-Lang, Danielle, and Karen Hébert. 2018. "Ecological Labor." Editors Forum: Theorizing the Contemporary. Society for Cultural Anthropology, July 26. https://culanth.org/fieldsights/ecological-labor.

Dooling, Sarah. 2009. "Ecological Gentrification: A Research Agenda Exploring Justice in the City." *International Journal of Urban and Regional Research* 33 (3): 621–39.

Dove, Michael. 1983. "Theories of Swidden Agriculture and the Political Economy of Ignorance." *Agroforestry Systems* 1:85–99.

du Bray, Margaret V., Rhian Stotts, Melissa Beresford, Amber Wutich, and Alexandra Brewis. 2019. "Does Ecosystem Services Valuation Reflect Local Cultural Valuations? Comparative Analysis of Resident Perspectives in Four Major Urban River Ecosystems." *Economic Anthropology* 6 (1): 21–33.

Duong, Kimberly, and Jean-Daniel M. Saphores. 2015. "Obstacles to Wastewater Reuse: An Overview." *Wiley Interdisciplinary Reviews: Water* 2 (3): 199–214.

Egge, Michael, and Idowu Ajibade. 2021. "A Community of Fear: Emotion and the Hydro-Social Cycle in East Porterville, California." *Journal of Political Ecology* 28: 266–85.

Elden, Stuart. 2013. "Secure the Volume: Vertical Geopolitics and the Depth of Power." *Political Geography* 34 (May): 35–51.

Elyachar, Julia. 2010. "Phatic Labor, Infrastructure, and the Question of Empowerment in Cairo." *American Ethnologist* 37 (3): 452–64.

Erie, Steven P. 2006. *Beyond Chinatown: The Metropolitan Water District, Growth, and the Environment in Southern California.* Stanford, CA: Stanford University Press.

Ernwein, Marion. 2020. "Bringing Urban Parks to Life: The More-Than-Human Politics of Urban Ecological Work." *Annals of the American Association of Geographers* 111 (2): 559–76.

Ernwein, Marion, Franklin Ginn, and James Palmer, eds. 2021. *The Work That Plants Do.* Bielefeld, Germany: Transcript.

Evans, Brad, and Reid, Julian. 2014. *Resilient Life: The Art of Living Dangerously.* Cambridge, UK: Polity.

Everett, G., J. E. Lamond, A. T. Morzillo, A. M. Matsler, and F. K. S. Chan. 2015. "Delivering Green Streets: An Exploration of Changing Perceptions and Behaviours over Time around Bioswales in Portland, Oregon." *Journal of Flood Risk Management* 11:5973–85.

Fairhead, James, and Melissa Leach. 1996. *Misreading the African Landscape: Society and Ecology in a Forest-Savanna Mosaic.* Cambridge: Cambridge University Press.

Federici, Silvia. 1975. *Wages Against Housework.* Bristol: Falling Wall Press.

Ferguson, James. 1999. *Expectations of Modernity: Myths and Meanings of Urban Life on the Zambian Copperbelt.* Berkeley: University of California Press.

Ferry, Elizabeth, and Mandana Limbert, eds. 2008. *Timely Assets: The Politics of Resources and Their Temporalities.* Santa Fe, NM: School for Advanced Research Press.

Finewood, Michael H. 2016. "Green Infrastructure, Grey Epistemologies, and the Urban Political Ecology of Pittsburgh's Water Governance." *Antipode* 48 (4): 1000–1021.

Finewood, Michael H., A. Marissa Matsler, and Joshua Zivkovich. 2019. "Green Infrastructure and the Hidden Politics of Urban Stormwater Governance in a Postindustrial City." *Annals of the American Association of Geographers* 109 (3): 909–25.

Fisher, Dana, Erika Svensden, and James Connolly. 2015. *Urban Environmental Stewardship and Civic Engagement.* New York: Routledge.

Fredericks, Rosalind. 2018. *Garbage Citizenship: Vital Infrastructures of Labor in Dakar, Senegal.* Durham, NC: Duke University Press.

Furlong, Kathryn, and Michelle Kooy. 2017. "Worlding Water Supply: Thinking Beyond the Network in Jakarta." *International Journal of Urban and Regional Research* 41 (6): 888–903.

Gaber, Nadia. 2021. "Blue Lines and Blues Infrastructures: Notes on Water, Race, and Space." *Environment and Planning D: Society and Space* 39 (6): 1073–91.

Gandy, Matthew. 2002a. "Between Borinquen and the Barrio: Environmental Justice and New York City's Puerto Rican Community, 1969–1972." *Antipode* 34 (4): 730–61.

———. 2002b. *Concrete and Clay: Reworking Nature in New York City.* Cambridge, MA: MIT Press.

———. 2004. "Rethinking Urban Metabolism: Water, Space and the Modern City." *City* 8 (3): 363–79.

———. 2006. "Riparian Anomie: Reflections on the Los Angeles River." *Landscape Research* 31 (2): 135–45.

———. 2014. *The Fabric of Space: Water, Modernity, and the Urban Imagination.* Cambridge, MA: MIT Press.

———. 2022. *Natura Urbana: Ecological Constellations in Urban Space.* Cambridge, MA: MIT Press.

Gidwani, Vinay. 2015. "The Work of Waste: Inside India's Infra-Economy." *Transactions of the Institute of British Geographers* 40 (4): 575–95.

Gleason, Victor. 1976. "Water Projects Go Underground." *Ecology Law Quarterly* 5 (4): 625–68.

Gleason, Victor E. 1977. "Los Angeles v. San Fernando: Ground Water Management in the Grand Tradition." *Hastings Constitutional Law Quarterly* 4 (4): 703–14.

Goh, Kian. 2021. *Form and Flow: The Spatial Politics of Urban Resilience and Climate Justice.* Cambridge, MA: MIT Press.

Gómez-Baggethun, Erik, Rudolf de Groot, Pedro L. Lomas, and Carlos Montes. 2010. "The History of Ecosystem Services in Economic Theory and Practice: From Early Notions to Markets and Payment Schemes." *Ecological Economics* 69 (6): 1209–18.

Gómez-Baggethun, Erik, and Manuel Ruiz-Pérez. 2011. "Economic Valuation and the Commodification of Ecosystem Services." *Progress in Physical Geography: Earth and Environment* 35 (5): 613–28.

Gonzalez, Silvia, Paul Ong, Gregory Pierce, and Ariana Hernandez. 2021. *Keeping the Lights and Water On: COVID-19 and Utility Debt in Los Angeles' Communities of Color.* Los Angeles: UCLA Luskin Center.

Graeber, David. 2001. *Toward an Anthropological Theory of Value: The False Coin of Our Own Dreams.* New York: Palgrave.

Graham, Stephen, and Simon Marvin. 2001. *Splintering Urbanism: Networked Infrastructures, Technological Mobilities and the Urban Condition.* London: Routledge.

Green, Dorothy. 2006. Interview by Jane Collings, May 9. Center for Oral History Research, University of California, Los Angeles. http://oralhistory .library.ucla.edu/.

———. 2008. *Managing Water: Avoiding Crisis in California.* Berkeley: University of California Press.

Greenleaf, Maron. 2020a. "Rubber and Carbon: Opportunity Costs, Incentives and Ecosystem Services in Acre, Brazil." *Development and Change* 51 (1): 51–72.

———. 2020b. "The Value of the Untenured Forest: Land Rights, Green Labor, and Forest Carbon in the Brazilian Amazon." *Journal of Peasant Studies* 47 (2): 286–305.

———. 2021. "Beneficiaries of Forest Carbon: Precarious Inclusion in the Brazilian Amazon." *American Anthropologist* 123 (2): 305–17.

Grove, Kevin. 2018. *Resilience.* New York: Routledge.

Gumprecht, Blake. 2001. *The Los Angeles River: Its Life, Death, and Possible Rebirth.* Baltimore, MD: Johns Hopkins University Press.

Günel, Gökçe. 2016. "The Infinity of Water: Climate Change Adaptation in the Arabian Peninsula." *Public Culture* 28 (2): 291–315.

———. 2019. *Spaceship in the Desert: Energy, Climate Change, and Urban Design in Abu Dhabi.* Durham, NC: Duke University Press.

Gupta, Akhil. 2018. "The Future in Ruins: Thoughts on the Temporality of Infrastructure." In *The Promise of Infrastructure,* edited by Nikhil Anand, Akhil Gupta, and Hannah Appel, 62–79. Durham, NC: Duke University Press.

Gusterson, Hugh. 1996. *Nuclear Rites: A Weapons Laboratory at the End of the Cold War.* Berkeley: University of California Press.

Harrison, Jill Lindsey. 2019. *From the Inside Out: The Fight for Environmental Justice within Government Agencies.* Urban and Industrial Environments. Cambridge, MA: MIT Press.

Hart, John. 1996. *Storm Over Mono: The Mono Lake Battle and the California Water Future.* Berkeley: University of California Press.

Harvey, David. 1996. *Justice, Nature, and the Geography of Difference.* Cambridge, MA: Blackwell.

Hastrup, Kirsten, and Rubow, Cecilie, eds. 2014. *Living with Environmental Change: Waterworlds.* New York: Routledge.

Haughton, Graham. 1998. "Private Profits–Public Drought: The Creation of a Crisis in Water Management for West Yorkshire." *Transactions of the Institute of British Geographers* 23 (4): 419–33.

Hawthorne, Camilla. 2019. "Black Matters Are Spatial Matters: Black Geographies for the Twenty-First Century." *Geography Compass* 13 (11): 1–13.

Hays, Samuel P. 1959. *Conservation and the Gospel of Efficiency: The Progressive Conservation Movement, 1890–1920*. Pittsburgh: University of Pittsburgh Press.

Hetherington, Kregg, ed. 2019. *Infrastructure, Environment, and Life in the Anthropocene*. Durham, NC: Duke University Press.

Hewitt, Alison. 2013. "UCLA Announces Plan to Tackle 'Grand Challenges,' Starting with Urban Sustainability." UCLA Newsroom, November 14. https://newsroom.ucla.edu/releases/grand-challenges-249212.

Heynen, Nik, Maria Kaika, and Erik Swyngedouw, eds. 2006. *In the Nature of Cities: Urban Political Ecology and the Politics of Urban Metabolism*. New York: Routledge.

Hoag, Colin. 2022. *The Fluvial Imagination: On Lesotho's Water-Export Economy*. Oakland: University of California Press.

Hoffman, Abraham. 1981. *Vision or Villainy: Origins of the Owens Valley-Los Angeles Water Controversy*. College Station: Texas A&M University Press.

Holifield, Ryan. 2001. "Defining Environmental Justice and Environmental Racism." *Urban Geography* 22 (1): 78–90.

Holling, C S. 1973. "Resilience and Stability of Ecological Systems." *Annual Review of Ecology and Systematics* 4 (1): 1–23.

Hommes, Lena, and Rutgerd Boelens. 2017. "Urbanizing Rural Waters: Rural-Urban Water Transfers and the Reconfiguration of Hydrosocial Territories in Lima." *Political Geography* 57: 71–80.

Hommes, Lena, Rutgerd Boelens, Leila M. Harris, and Gert Jan Veldwisch. 2019. "Rural–Urban Water Struggles: Urbanizing Hydrosocial Territories and Evolving Connections, Discourses and Identities." *Water International* 44 (2): 81–94.

Hommes, Lena, Jaime Hoogesteger, and Rutgerd Boelens. 2022. "(Re)Making Hydrosocial Territories: Materializing and Contesting Imaginaries and Subjectivities through Hydraulic Infrastructure." *Political Geography* 97 (August): 102698.

Hondagneu-Sotelo, Pierrette. 2014. *Paradise Transplanted: Migration and the Making of California Gardens*. Oakland: University of California Press.

Huang, Xingying, Samantha Stevenson, and Alex D. Hall. 2020. "Future Warming and Intensification of Precipitation Extremes: A 'Double Whammy' Leading to Increasing Flood Risk in California." *Geophysical Research Letters* 47 (16): e2020GL088679.

Huff, Amber. 2021. "Frictitious Commodities: Virtuality, Virtue and Value in the Carbon Economy of Repair." *Environment and Planning E: Nature and Space* (June): 25148486211015056.

Hughes, Sara, Stephanie Pincetl, and Christopher Boone. 2013. "Triple Exposure: Regulatory, Climatic, and Political Drivers of Water Management Changes in the City of Los Angeles." *Cities* 32:51–59.

Hundley, Norris. 2001. *The Great Thirst: Californians and Water, a History.* Berkeley: University of California Press.

Hundley, Norris, Donald C. Jackson, and Jean Patterson. 2016. *Heavy Ground: William Mulholland and the St. Francis Dam Disaster.* Oakland: University of California Press.

Ingham, Alan. 1980. "Residential Greywater Management." *California Department of Water Resources Bulletin.*

Isenhour, Cindy. 2011. "How the Grass Became Greener: On Urban Imaginings and Practices of Sustainable Living in Sweden." *City & Society* 23 (2): 117–34.

Isenhour, Cindy, Gary McDonough, and Melissa Checker, eds. 2015. *Sustainability in the City: Myth and Practice.* Cambridge: Cambridge University Press.

Jackson, Sue, and Lisa R. Palmer. 2015. "Reconceptualizing Ecosystem Services: Possibilities for Cultivating and Valuing the Ethics and Practices of Care." *Progress in Human Geography* 39 (2): 122–45.

James, Ian. 2024. "Honeywell Will Fund Cleanup of Contaminated Groundwater in San Fernando Valley, EPA Says." *Los Angeles Times*, October 1, 2024.

Jensen, Casper Bruun, and Atsuro Morita. 2015. "Infrastructures as Ontological Experiments." *Engaging Science, Technology, and Society* 1 (November): 81–87.

Johnson, Leigh, Michael Mikulewicz, Patrick Bigger, Ritodhi Chakraborty, Abby Cunniff, P. Joshua Griffin, Vincent Guermond, Nicole Lambrou, Megan Mills-Novoa, Benjamin Neimark, Sara Nelson, Costanza Rampini, Pasang Sherpa, and Gregory Simon. 2023. "Intervention: The Invisible Labor of Climate Change Adaptation." *Global Environmental Change* 83:102769.

Joslin, Audrey. 2020. "Translating Water Fund Payments for Ecosystem Services in the Ecuadorian Andes." *Development and Change* 51 (1): 94–116.

Kahrl, William M. 1982. *Water and Power: The Conflict over Los Angeles' Water Supply in the Owens Valley.* Berkeley: University of California Press.

Kaika, M., and E. Swyngedouw. 2002. "Fetishizing the Modern City: The Phantasmagoria of Urban Technological Networks." *International Journal of Urban and Regional Research* 24 (1): 120–38.

Kaika, Maria. 2003. "Constructing Scarcity and Sensationalising Water Politics: 170 Days That Shook Athens." *Antipode* 35 (5): 919–54.

———. 2005. *City of Flows: Modernity, Nature, and the City.* New York: Routledge.

———. 2006. "Dams as Symbols of Modernization: The Urbanization of Nature Between Geographical Imagination and Materiality." *Annals of the Association of American Geographers* 96 (2): 276–301.

Kama, Kärg. 2020. "Resource-Making Controversies: Knowledge, Anticipatory Politics and Economization of Unconventional Fossil Fuels." *Progress in Human Geography* 44 (2): 333–56.

———. 2021. "Temporalities of (Un) Making a Resource: Oil Shales Between Presence and Absence." In *The Routledge Handbook of Critical Resource Geography*, edited by Matthew Himley, Elizabeth Havice, and Gabriela Valdivia, 57–67. New York: Routledge.

Karvoven, Andrew. 2011. *The Politics of Urban Runoff: Nature, Technology, and the Sustainable City*. Cambridge, MA: MIT Press.

Kim, Nadia Y. 2021. *Refusing Death: Immigrant Women and the Fight for Environmental Justice in LA*. Stanford, CA: Stanford University Press.

Kinder, Kimberley. 2016. *DIY Detroit: Making Do in a City without Services*. Minneapolis: University of Minnesota Press.

Kiparsky, Michael, Kathleen Miller, William Blomquist, Annapurna Holtzapple, and Anita Milman. 2021. "Groundwater Recharge to Address Seawater Intrusion and Supply in an Urban Coastal Aquifer." *Case Studies in the Environment* 5 (1): 1–10.

Kluckhohn, Clyde. 1961. "The Study of Values." In *Values in America*, edited by Donald Barrett, 17–45. South Bend, IN: University of Notre Dame Press.

Kneas, David. 2018. "Emergence and Aftermath: The (Un)Becoming of Resources and Identities in Northwestern Ecuador." *American Anthropologist* 120 (4): 752–64.

———. 2020. "Translating Tierra Baldía: Land, Landscape, and the Becoming of Frontier Terrain in Northwestern Ecuador." *Ethnos* 85 (1): 33–53.

Knox, Hannah. 2020. *Thinking Like a Climate: Governing a City in Times of Environmental Change*. Durham, NC: Duke University Press.

Koslov, Liz. 2016. "The Case for Retreat." *Public Culture* 28 (2): 359–88.

Krause, Franz. 2013. "Seasons as Rhythms on the Kemi River in Finnish Lapland." *Ethnos* 78 (1): 23–46.

———. 2022a. "Rhythms of Wet and Dry: Temporalising the Land-Water Nexus." *Geoforum* 131: 252–59.

———. 2022b. "The Tempo of Solid Fluids: On River Ice, Permafrost, and Other Melting Matter in the Mackenzie Delta." *Theory, Culture & Society* 2: 31–52.

Kroepsch, Adrianne C., and Katherine R. Clifford. 2022. "On Environments of Not Knowing: How Some Environmental Spaces and Circulations Are Made Inscrutable." *Geoforum* 132 (June): 171–81.

Kurtiç, Ekin. 2023. "Infrastructural Decay." *Cultural Anthropology* 38 (1): 142–70.

Lamond, Jessica, and Glyn Everett. 2023. "Willing to Have, Willing to Help, or Ready to Own —Determinants of Variants of Stewardship Social Practices Around Blue-Green Infrastructure in Dense Urban Communities." *Frontiers in Water* 5:1048494.

Lane, D. A. 1934. "Increasing Storage by Water Spreading." *Journal of the American Water Works Association* 26 (4): 421–29.

Larkin, Brian. 2008. *Signal and Noise: Media, Infrastructure, and Urban Culture in Nigeria*. Durham, NC: Duke University Press.

Latour, Bruno. 1987. *Science in Action: How to Follow Scientists and Engineers through Society*. Cambridge, MA: Harvard University Press.

———. 2004. *Politics of Nature: How to Bring the Sciences into Democracy*. Cambridge, MA: Harvard University Press.

Lefebvre, Henri. 2004. *Rhythmanalysis: Space, Time, and Everyday Life*. London: Continuum.

Lejano, Raul P., and Jonathon E. Ericson. 2005. "Tragedy of the Temporal Commons: Soil-Bound Lead and the Anachronicity of Risk." *Journal of Environmental Planning and Management* 48 (2): 301–20.

Leopold, Luna. 1968. *Hydrology for Urban Land Planning: A Guidebook on the Hydrologic Effects of Urban Land Use*. Reston, VA: United States Geological Survey.

Leslie, Jacques. 2014. "Los Angeles, City of Water." *New York Times*, December 6, 2014.

Levin, Myron. 1985a. "Plans for Sewers Aimed at Curbing Pollution of Wells." *Los Angeles Times*, August 26, 1985.

———. 1985b. "Polluted Wells in the East Valley: DWP Pleads for a Rush on Cleanup." *Los Angeles Times*, September 27, 1985.

Ley, Lukas. 2021. *Building on Borrowed Time: Rising Seas and Failing Infrastructure in Semarang*. Minneapolis: University of Minnesota Press.

Limbert, Mandana. 2010. *In the Time of Oil: Piety, Memory & Social Life in an Omani Town*. Stanford, CA: Stanford University Press.

Lock, Will. 2023. "Producing Nature-Based Solutions: Infrastructural Nature and Agrarian Change in San Martín, Peru." *Journal of Peasant Studies* 50 (6): 2361–80.

Lopez, Robert. 2022. "Massive Santa Monica Bay Sewage Spill Likely Caused by Human Error, Equipment Failure." *Los Angeles Times*, February 11, 2022.

Los Angeles County Flood Control Act. 1915. Statutes of 1915, Ch. 755.

Los Angeles Department of Water and Power (LADWP). 2015. "Groundwater System Improvement Study." https://www.ladwp.com/community /construction-projects/other/groundwater-remediation.

———. 2016. "Stormwater Capture Master Plan." https://www.ladwp.com/sites /default/files/2023-08/LADWPStormwaterCaptureMasterPlan_MainReport _101615.pdf.

———. 2021. "Urban Water Management Plan 2020." https://www.ladwp.com /sites/default/files/documents/LADWP_2020_UWMP_Web.pdf.

Los Angeles Homeless Services Authority. 2022. "Homeless Count by City Council District." https://www.lahsa.org/data?id=52-homeless-count-by-city -of-la-council-district-2015-2022.

Los Angeles Times. 1899. "Arbor et Aqua." July 21, 1899.

Los Angeles Times. 1931. "Water Salvage Land Available." December 14, 1931.

Los Angeles Times. 1968. "Water Spreading Is Obsolete, Study Says." October 6, 1968.

Los Angeles Waterkeeper. 2023. *Changing the Course? What's Worked, What Hasn't, and What's Next for the SCWP: An Assessment of the First Three Rounds of the Safe Clean Water Program Regional Funding Program*. https://drive.google.com/file/d/1kEIlOeEPdMCltzan6jLpzNZt4z4YE_8g/view

Lu, Zita, Anditya Rahardianto, J. R. DeShazo, Michael Stenstrom, and Yoram Cohen. 2013. "Critical Review: Regulatory Incentives and Impediments for Onsite Graywater Reuse in the United States." *Water Environment Research* 85 (7): 650–62..

Lubas, Ken. 1978. "Rain-Ravaged Graves Not in Cemetery: Course Count of 30 May Higher, Public Works Official Says." *Los Angeles Times*, March 2, 1978.

Maida, Carl. 2011. "Participatory Action Research and Urban Environmental Justice: The Pacoima CARE Project." In *Environmental Anthropology Today*, edited by Helen Kopnina and Eleanor Shoreman-Ouimet, 161–75. New York: Routledge.

Majuru, Batsirai, Marc Suhrcke, and Paul R. Hunter. 2016. "How Do Households Respond to Unreliable Water Supplies? A Systematic Review." *International Journal of Environmental Research and Public Health* 13 (12): 1222.

Malpas, Jeff, ed. 2011. *The Place of Landscape: Concepts, Contexts, Studies*. Cambridge, MA: MIT Press.

Mathews, Andrew S., and Jessica Barnes. 2016. "Prognosis: Visions of Environmental Futures." *Journal of the Royal Anthropological Institute* 22 (S1): 9–26.

Matsler, A. Marissa. 2019. "Making 'Green' Fit in a 'Grey' Accounting System: The Institutional Knowledge System Challenges of Valuing Urban Nature as Infrastructural Assets." *Environmental Science & Policy* 99 (September): 160–68.

Maurer, Megan. 2020. "Nourishing Environments, Caring Cities: Gardening and Social Reproduction of the Urban Environment in Deindustrial Michigan." *City & Society* 32 (3): 716–37.

———. 2024. "Vegetal Agency in Street Tree Stewardship Practices: People-Plant Involutions Within Urban Green Infrastructure in New York City." *Journal of Ethnobiology* 44 (1): 56–68.

McElwee, Pamela D. 2016. *Forests Are Gold: Trees, People, and Environmental Rule in Vietnam*. Seattle: University of Washington Press.

———. 2017. "The Metrics of Making Ecosystem Services." *Environment and Society* 8 (1): 96–124.

McGee, William J. 1909. "Water as a Resource." *Annals of the American Academy of Political and Social Science* 33: 521–34.

McHarg, Ian. 1969. *Design with Nature*. New York: American Museum of Natural History.

McKittrick, Katherine. 2013. "Plantation Futures." *Small Axe: A Caribbean Journal of Criticism* 17 (3): 1–15.

Meehan, Katharine, Kerri Jean Ormerod, and Sarah A Moore. 2013. "Remaking Waste as Water: The Governance of Recycled Effluent for Potable Water Supply." *Water Alternatives* 6 (1): 67–85.

Meehan, Katie, Wendy Jepson, Leila M. Harris, Amber Wutich, Melissa Beresford, Amanda Fencl, Jonathan London, et al. 2020. "Exposing the Myths of Household Water Insecurity in the Global North: A Critical Review." *Wiley Interdisciplinary Reviews: Water* 7 (6): 1–20.

Meehan, Katie, Jason Jurjevich, Alison Griswold, Nicholas Chun, and Justin Sherrill. 2021. *Plumbing Poverty in U.S. Cities: A Report on Gaps and Trends in Household Water Access, 2000 to 2017.* London: King's College London.

Meehan, Katie M. 2014. "Tool-Power: Water Infrastructure as Wellsprings of State Power." *Geoforum* 57:215–24.

Meilinger, Valentin, and Jochen Monstadt. 2022a. "From the Sanitary City to the Circular City? Technopolitics of Wastewater Restructuring in Los Angeles, California." *International Journal of Urban and Regional Research* 46 (2): 182–201.

———. 2022b. "The Material Politics of Integrated Urban Stormwater Management in Los Angeles, California." *Local Environment* 27 (7): 847–62.

———. 2023. "Infrastructuring Gardens: The Material Politics of Outdoor Water Conservation in Los Angeles." *Annals of the American Association of Geographers* 113 (1): 206–24.

Melosi, Marvin. 2000. *The Sanitary City: Urban Infrastructure in America from Colonial Times to the Present.* Baltimore, MD: Johns Hopkins University Press.

Méndez-Barrientos, Linda E., Amanda L. Fencl, Cassandra L. Workman, and Sameer H. Shah. 2023. "Race, Citizenship, and Belonging in the Pursuit of Water and Climate Justice in California." *Environment and Planning E: Nature and Space* 6 (3): 1614–35.

Mendoza, AnMarie. 2019. "The Aqueduct Between Us-Inserting and Asserting an Indigenous California Indian Perspective about Los Angeles Water" (Master's Thesis, University of California Los Angeles).

Millington, Nate, and Suraya Scheba. 2020. "Day Zero and the Infrastructures of Climate Change: Water Governance, Inequality, and Infrastructural Politics in Cape Town's Water Crisis." *International Journal of Urban and Regional Research* 45 (1): 116–32.

Mills-Novoa, Megan. 2023. "What Happens after Climate Change Adaptation Projects End: A Community-Based Approach to Ex-post Assessment of Adaptation Projects." *Global Environmental Change* 80:102655.

Mini, Caroline, Terri S. Hogue, and Stephanie Pincetl. 2014. "Estimation of Residential Outdoor Water Use in Los Angeles, California." *Landscape and Urban Planning* 127:124–35.

Mitchell, Don. 1996. *The Lie of the Land: Migrant Workers and the California Landscape*. Minneapolis: University of Minnesota Press.

———. 2002. "Cultural Landscapes: The Dialectical Landscape—Recent Landscape Research in Human Geography." *Progress in Human Geography* 26 (3): 381–89.

Mitchelson, A. T. 1930. "Storage of Water Underground by Spreading Over Absorptive Areas." *California State Department of Water Resources Bulletin* 32:49–56.

Mohai, Paul, David Pellow, and J. Timmons Roberts. 2009. "Environmental Justice." *Annual Reviews in Environmental Resources* 34:405–30.

Molle, François. 2009. "River-Basin Planning and Management: The Social Life of a Concept." *Geoforum* 40:484–94.

Molyneux, Maxine. 1979. "Beyond the Domestic Labor Debate." *New Left Review* 116:3–27.

Moore, Amelia. 2019. "Selling Anthropocene Space: Situated Adventures in Sustainable Tourism." *Journal of Sustainable Tourism* 27 (4): 436–51.

Moore, Jason W. 2015. *Capitalism in the Web of Life: Ecology and the Accumulation of Capital*. New York: Verso.

Morgan, Ruth. 2020. "The Allure of Climate and Water Independence: Desalination Projects in Perth and San Diego." *Journal of Urban History* 46 (1): 113–28.

Muehlebach, Andrea Karin. 2012. *The Moral Neoliberal: Welfare and Citizenship in Italy*. Chicago Studies in Practices of Meaning. Chicago: University of Chicago Press.

Nadeau, Remi. 1960. *Los Angeles: From Mission to Modern City*. London: Longmans, Green.

Nader, Laura. 1974. "Up the Anthropologist: Perspectives Gained from Studying Up." In *Reinventing Anthropology*, edited by Dell Hymes, 284–311. New York: Vintage Books.

Needham, Andrew. 2014. *Power Lines: Phoenix and the Making of the Modern Southwest*. Princeton, NJ: Princeton University Press.

Nelson, Sara H. 2015. "Beyond *The Limits to Growth*: Ecology and the Neoliberal Counterrevolution." *Antipode* 47 (2): 461–80.

Nelson, Sara H., and Patrick Bigger. 2022. "Infrastructural Nature." *Progress in Human Geography* 46 (1): 86–107.

Nelson, Sara Holiday, Leah L. Bremer, and Kelly Meza Prado. 2020. "The Political Life of Natural Infrastructure: Water Funds and Alternative Histories of Payments for Ecosystem Services in Valle Del Cauca, Colombia." *Development and Change* 51 (1): 26–50.

Nevarez, Leonard. 1996. "Just Wait until There's a Drought: Mediating Environmental Crises for Urban Growth." *Antipode* 28 (3): 246–72.

Newman, Andrew. 2011. "Contested Ecologies: Environmental Activism and Urban Space in Immigrant Paris." *City & Society* 23 (2): 192–209.

———. 2015. *Landscape of Discontent: Urban Sustainability in Immigrant Paris*. Minneapolis.: University of Minnesota Press.

Nost, Eric. 2022. "'The Tool Didn't Make Decisions for Us': Metrics and the Performance of Accountability in Environmental Governance." *Science as Culture* 33(1): 97–120.

Nost, Eric, and Jenny Goldstein. 2022. "A Political Ecology of Data." *Environment and Planning E: Nature and Space* 5 (1): 3–17.

Nucho, Joanne. 2017. *Everyday Sectarianism in Urban Lebanon: Infrastructures, Public Services, and Power*. Princeton, NJ: Princeton University Press.

Ojani, Chakad. 2023. "Experimenting with Fog: Environmental Infrastructures, Infrastructuring Environments, and the Infrastructure of Infrastructure." *Environment and Planning E: Nature and Space* 6 (1): 24–41.

O'Leary, Heather. 2016. "Between Stagnancy and Affluence: Reinterpreting Water Poverty and Domestic Flows in Delhi, India." *Society & Natural Resources* 29 (6): 639–53.

Olson, Valerie. 2018. *Into the Extreme: U.S. Environmental Systems and Politics beyond Earth*. Minneapolis: University of Minnesota Press.

O'Neill, Brian. 2023. "Water for Whom? Desalination and the Cooptation of the Environmental Justice Frame in Southern California." *Environment and Planning E: Nature and Space* 6 (2): 1366–90.

O'Neill, Karen. 2006. *Rivers by Design: State Power and the Origins of U.S. Flood Control*. Durham, NC: Duke University Press.

O'Reilly, Jessica. 2016. "Sensing the Ice: Field Science, Models, and Expert Intimacy with Knowledge." *Journal of the Royal Anthropological Institute* 22 (S1): 27–45.

Ormerod, Kerri Jean. 2016. "Illuminating Elimination: Public Perception and the Production of Potable Water Reuse." *Wiley Interdisciplinary Reviews: Water* 3:537–47.

———. 2019. "Toilet Power: Potable Water Reuse and the Situated Meaning of Sustainability in the Southwestern United States." *Journal of Political Ecology* 26 (1): 633–51.

Orsi, Jared. 2004. *Hazardous Metropolis: Flooding and Urban Ecology in Los Angeles*. Berkeley: University of California Press.

Osborne, Tracey, and Elizabeth Shapiro-Garza. 2018. "Embedding Carbon Markets: Complicating Commodification of Ecosystem Services in Mexico's Forests." *Annals of the American Association of Geographers* 108 (1): 88–105.

Owens, Caitlin, James Queally, and Emily Alpert Reyes. 2014. "UCLA-Area Water Main Break Spews Millions of Gallons." *Los Angeles Times*, July 29, 2014.

Paprocki, Kasia. 2021. *Threatening Dystopias: The Global Politics of Climate Change Adaptation in Bangladesh*. Ithaca, NY: Cornell University Press.

Park, Lisa Sun-Hee, and David Naguib Pellow. 2011. *The Slums of Aspen: Immigrants vs. the Environment in America's Eden*. New York: New York University Press.

Park, Mi-Hyun, Michael Stenstrom, and Stephanie Pincetl. 2009. "Water Quality Improvement Policies: Lessons Learned from the Implementation of Proposition O in Los Angeles, California." *Environmental Management* 43:514–22.

Parreñas, Juno Salazar. 2018. *Decolonizing Extinction: The Work of Care in Orangutan Rehabilitation*. Durham, NC: Duke University Press.

Petryna, Adriana. 2022. *Horizon Work: At the Edges of Knowledge in an Age of Runaway Climate Change*. Princeton, NJ: Princeton University Press.

Piechota, Tom, and Suzanne Dallman. 2010. *Stormwater: Asset, Not Liability*. Los Angeles: Council for Watershed Health.

Pincetl, Stephanie, Erik Porse, Kathryn Mika, Elizaveta Litvak, Kimberly Manago, Terri Hogue, Thomas Gillespie, Diane Pataki, and Mark Gold. 2019. "Adapting Urban Water Systems to Manage Scarcity in the 21st Century." *Environmental Management* 63:293–308.

Piper, Karen. 2006. *Left in the Dust: How Race and Politics Created a Human and Environmental Tragedy in L.A.* New York: Macmillan.

Pompeii, Brian. 2020. "The Social Production of the Great California Drought, 2012–2017." *Yearbook of the Association of Pacific Coast Geographers* 82:15–37.

Ponce, Mary Helen. (1992) 2006. *Hoyt Street: An Autobiography*. Albuquerque: University of New Mexico Press.

Powell, Dana E. 2018. *Landscapes of Power: Politics of Energy in the Navajo Nation*. Durham, NC: Duke University Press.

Powis, Anthony. 2021. "The Relational Materiality of Groundwater." *GeoHumanities* 7 (1): 89–112.

Pulido, Laura. 2000. "Rethinking Environmental Racism: White Privilege and Urban Development in Southern California." *Annals of the Association of American Geographers* 90 (1): 12–40.

———. 2016. "Flint, Environmental Racism, and Racial Capitalism." *Capitalism Nature Socialism* 27 (3): 1–16.

Qiao, Xiu-Juan, Li Liu, Anders Kristoffersson, and Thomas B Randrup. 2019. "Governance Factors of Sustainable Stormwater Management: A Study of Case Cities in China and Sweden." *Journal of Environmental Management* 248:109249.

Rademacher, Anne. 2011. *Reigning the River: Urban Ecologies and Political Transformation in Kathmandu*. Durham, NC: Duke University Press.

Rademacher, Anne, and Kalyanakrishnan Sivaramakrishnan, eds. 2013. *Ecologies of Urbanism in India: Metropolitan Civility and Sustainability*. Hong Kong: Hong Kong University Press.

Radonic, Lucero. 2019a. "Becoming with Rainwater: A Study of Hydrosocial Relations and Subjectivity in a Desert City." *Economic Anthropology* 6 (2): 291–303.

———. 2019b. "Re-Conceptualising Water Conservation: Rainwater Harvesting in the Desert of the Southwestern United States." *Water Alternatives* 12 (2): 699–714.

Randle, Sayd. 2020. "Ordinary Disasters: On Unexceptional Flooding in LA's San Fernando Valley." *Michigan Quarterly Review* 2:237–48.

———. 2021. "Missing Power: Nostalgia and Disillusionment among Southern California Water Engineers." *Critique of Anthropology* 41 (3): 267–83.

———. 2022a. "Holding Water for the City: Emergent Geographies of Storage and the Urbanization of Nature." *Environment and Planning E: Nature and Space* 5 (4): 2283–2306.

———. 2022b. "On Aqueducts and Anxiety: Water Infrastructure, Ruination, and a Region-Scaled Anthropocene Imaginary." *GeoHumanities* 8 (1): 33–52.

Redfield, Peter. 2000. *Space in the Tropics: From Convicts to Rockets in French Guiana*. Berkeley: University of California Press.

Reibel, Michael, Madelyn Glickfeld, and Peter Roquemore. 2021. "Disadvantaged Communities and Drinking Water: A Case Study of Los Angeles County." *GeoJournal* 86 (3): 1337–54.

Reisner, Marc. (1986) 1993. *Cadillac Desert: The American West and Its Disappearing Water*. New York: Penguin Books.

Richardson, Tanya, and Gisa Weszkalnys. 2014. "Resource Materialities." *Anthropological Quarterly* 87 (1): 5–30.

Riedman, Elizabeth. 2021. "Othermothering in Detroit, MI: Understanding Race and Gender Inequalities in Green Stormwater Infrastructure Labor." *Journal of Environmental Policy and Planning* 23 (5): 616–27.

Roberts, Jason. 2013. "'What Are We Protecting Out Here?' A Political Ecology of Forest, Fire, and Fuels Management in Utah's Wildland-Urban Interface." *Capitalism Nature Socialism* 24 (2): 58–76.

Robertson, Morgan M. 2006. "The Nature That Capital Can See: Science, State, and Market in the Commodification of Ecosystem Services." *Environment and Planning D: Society and Space* 24 (3): 367–87.

———. 2007. "Discovering Price in All the Wrong Places: The Work of Commodity Definition and Price under Neoliberal Environmental Policy." *Antipode* 39 (3): 500–526.

———. 2012. "Measurement and Alienation: Making a World of Ecosystem Services." *Transactions of the Institute of British Geographers* 37 (3): 386–401.

Robbins, Paul. 2007. *Lawn People: How Grasses, Weeds, and Chemicals Make Us Who We Are*. Philadelphia: Temple University Press.

Roderick, Kevin. 2001. *The San Fernando Valley: America's Suburb*. Los Angeles: Los Angeles Times.

Roesner, Larry, Yaling Qian, Melanie Criswell, Mary Stromberger, and Stephen Klein. 2006. *Long-Term Effects of Landscape Irrigation Using Household Graywater—Literature Review and Synthesis*. Alexandria, VA: Water Environment Research Foundation.

Rome, Adam. 2001. *The Bulldozer in the Countryside: Suburban Sprawl and the Rise of American Environmentalism*. Cambridge: Cambridge University Press.

Rosaldo, Michelle, and Louise Lamphere, eds. 1974. *Woman, Culture, and Society*. Stanford, CA: Stanford University Press.

Rosenbaum, Susanna. 2017. *Domestic Economies: Women, Work, and the American Dream*. Durham, NC: Duke University Press.

Ryan, Ruth. 1985. "Toxic Waste: Past Misuse Haunts the Present." *Los Angeles Times*, November 3, 1985.

Saguin, Kristian. 2022. *Urban Ecologies on the Edge: Making Manila's Resource Frontier*. Oakland: University of California Press.

Sahagun, Louis. 2023. "L.A.'s New Water War: Keeping Supply from Mono Lake Flowing as Critics Want it Cut Off." *Los Angeles Times*, February 19, 2023.

Salinas, Ivan. 2019. "Rushing Waters: A Mural that Celebrates the History of Pacoima." *Daily Sundial*, November 20. https://sundial.csun.edu/155630 /arts-entertainment/rushing-waters-a-mural-that-celebrates-the-history-of -pacoima/.

Scaramelli, Caterina. 2019. "The Delta Is Dead: Moral Ecologies of Infrastructure in Turkey." *Cultural Anthropology* 34 (3): 388–416.

———. 2021. *How to Make a Wetland: Water and Moral Ecology in Turkey*. Stanford, CA: Stanford University Press.

Scavo, Jordan. 2013. "Water Politics and the San Fernando Valley: The Role of Water Rights." *Southern California Quarterly* 92 (2): 93–116.

Schmidt, Charles. 2008. "The Yuck Factor When Disgust Meets Discovery." *Environmental Health Perspectives* 116 (2): A524–27.

Scott, Allen. 1996. "High Technology Industrial Development in the San Fernando Valley and Ventura County." In *The City*, edited by Allen Scott and Edward Soja, 276–310. Berkeley: University of California Press.

Scott, James C. 2006. "High Modernist Social Engineering: The Case of the Tennessee Valley Authority." In *Experiencing the State*, edited by Llyod Rudolph and John Jacobsen, 3–52. New Delhi: Oxford University Press.

Shamasunder, Bhavna, Raquel Mason, Luke Ippoliti, and Laura Robledo. 2015. "Growing Together: Poverty Alleviation, Community Building, and Environmental Justice through Home Gardens in Pacoima, Los Angeles." *Environmental Justice* 8 (3): 72–77.

Shapiro-Garza, Elizabeth, Pamela McElwee, Gert Van Hecken, and Esteve Corbera. 2020. "Beyond Market Logics: Payments for Ecosystem Services as Alternative Development Practices in the Global South." *Development and Change* 51 (1): 3–25.

Sharpsteen, Bill. 2010. *Dirty Water: One Man's Fight to Clean up One of the World's Most Polluted Bays*. Berkeley: University of California Press.

Shreenath, Shreyas. 2023. "(Un)making the Manual Scavenger: Caste, Contract, and Ecological Uncertainty in Bengalaru, India." *American Ethnologist* 50 (3): 491–505.

Sides, Josh. 2003. *L.A. City Limits: African American Los Angeles from the Great Depression to the Present*. Berkeley: University of California Press.

Simon, Gregory. 2017. *Flame and Fortune in the American West: Urban Development, Environmental Change, and the Great Oakland Hills Fire*. Oakland: University of California Press.

Simone, AbdouMaliq. 2004. "People as Infrastructure: Intersecting Fragments in Johannesburg." *Public Culture* 16 (3): 407–29.

Singh, Neera M. 2013. "The Affective Labor of Growing Forests and the Becoming of Environmental Subjects: Rethinking Environmentality in Odisha, India." *Geoforum* 47:189–98.

Sister, Chona, Jennifer Wolch, and John Wilson. 2010. "Got Green? Addressing Environmental Justice in Park Provision." *GeoJournal* 75 (3): 229–48.

Sizek, Julia. 2023. "Regulatory Alchemy: How the Water Cycle Becomes Capital in the California Desert." *Antipode* 55 (6): 1898–1918.

Sklar, Anna. 2008. *Brown Acres: An Intimate History of the Los Angeles Sewers*. Los Angeles: Angel City Press.

Smith, Hayley. 2023. "California Is Letting Billions of Gallons of Stormwater Wash Out to Sea Each Year, Report Finds." *Los Angeles Times*, March 1, 2023.

Sofoulis, Zoe. 2005. "Big Water, Everyday Water: A Sociotechnical Perspective." *Continuum: Journal of Media & Cultural Studies* 19 (4): 445–63.

Sony, R. K., and Siddhartha Krishnan. 2023. "Riverine Relations, Affective Labor and Changing Environmental Subjectivity in Kerala, South India." *Geoforum* 140:103701.

Souleles, Daniel. 2018. "How to Study People Who Do Not Want to Be Studied: Practical Reflections on Studying Up." *Political and Legal Anthropology Review* 41 (S1): 51–68.

———. 2021. "How to Think about People Who Don't Want to Be Studied: Further Reflections on Studying Up." *Critique of Anthropology* 41 (3): 206–26.

Stamatopoulou-Robbins, Sophia. 2020. *Waste Siege: The Life of Infrastructure in Palestine*. Stanford Studies in Middle Eastern and Islamic Societies and Cultures. Stanford, CA: Stanford University Press.

Stamatopoulou-Robbins, Sophia C. 2021. "Failure to Build: Sewage and the Choppy Temporality of Infrastructure in Palestine." *Environment and Planning E: Nature and Space* 4 (1): 28–42.

Stein, Mark. 1983. "Wells Safe Now, but Future Uncertain, City Engineers Say." *Los Angeles Times*, January 13, 1983, V1.

Stilgoe, John R. 2018. *What Is Landscape?* Cambridge, MA: MIT Press.

Stoetzer, Bettina. 2022. *Ruderal City: Ecologies of Migration, Race, and Urban Nature in Berlin*. Durham, NC: Duke University Press.

Stokes, Kathleen, and Alejandro De Coss-Corzo. 2023. "Doing the Work: Locating Labour in Infrastructural Geography." *Progress in Human Geography* (May): 03091325231174186.

Strang, Veronica. 2004. *The Meaning of Water*. Oxford; New York: Berg.

———. 2009. *Gardening the World: Agency, Identity, and the Ownership of Water*. London: Beghahn.

Suarez, Daniel Chiu. 2023. "Mainstreaming Ecosystem Services: The Hard Work of Realigning Biodiversity Conservation." *Environment and Planning E: Nature and Space* 6 (2): 1299–1321.

Sullivan, Sian. 2018. "Making Nature Investable: From Legibility to Leverage-ability in Fabricating 'Nature' as 'Natural Capital.'" *Science and Technology Studies* 31 (3): 47–76.

Swyngedouw, Erik. 2005. "Governance Innovation and the Citizen: The Janus Face of Governance-beyond-the-State." *Urban Studies* 42 (11): 1991–2006.

Taguchi, T. L. 2003. "Whose Space Is It Anyway: Protecting the Public Interest in Allocating Storage Space in California's Groundwater Basins." *Southwestern University Law Review* 32 (1): 117–50.

Taylor, Dorceta E. 2014. *Toxic Communities: Environmental Racism, Industrial Pollution, and Residential Mobility*. New York: New York University Press.

Tozer, Laura, Harriet Bulkeley, Bernadett Kiss, Andrés Luque-Ayala, Yuliya Voytenko Palgan, Kes McCormick, and Christine Wamsler. 2023. "Nature for Resilience? The Politics of Governing Urban Nature." *Annals of the American Association of Geographers* 113 (3): 599–615.

Truelove, Yaffa. 2019. "Gray Zones: The Everyday Practices and Governance of Water beyond the Network." *Annals of the American Association of Geographers* 109 (6): 1758–74.

———. 2021. "Who Is the State? Infrastructural Power and Everyday Water Governance in Delhi." *Environment and Planning C: Politics and Space* 39 (2): 282–99.

Tsing, Anna Lowenhaupt. 2015. *The Mushroom at the End of the World: On the Possibility of Life in Capitalist Ruins*. Princeton, NJ: Princeton University Press.

Turnhout, Esther, Tamara Metze, Carina Wyborn, Nicole Klenk, Elena Louder, James C Arnott, Katharine J Mach, and Gabrielle Wong-Parodi. 2019. "The Politics of Co-Production: Participation, Power, and Transformation." *Current Opinion in Environmental Sustainability* 42:15–21.

US Bureau of Reclamation. 2016. *Los Angeles Basin Study: The Future of Stormwater Conservation*. Denver, CO: US Department of the Interior, Bureau of Reclamation Technical Service Center.

US EPA (United States Environmental Protection Agency). 2019. "What Is Green Infrastructure?" May 29. https://www.epa.gov/green-infrastructure /what-green-infrastructure.

Van Bueren, Thad M. 2002. "Struggling with Class Relations at a Los Angeles Aqueduct Construction Camp." *Historical Archaeology* 36:28–43.

Vartabedian, Ralph. 2022. "Red Tape Ensnares L.A.'s Plan to Capture More Storm Water." *Los Angeles Times*, March 4, 2022.

Vaughn, Sarah E. 2022. *Engineering Vulnerability: In Pursuit of Climate Adaptation*. Durham, NC: Duke University Press.

Villaraigosa, Antonio, and the Los Angeles Department of Water and Power. 2008. *Securing LA's Water Supply*. Los Angeles: City of Los Angeles. https://www.ladwp.com/sites/default/files/documents/2008_Water_Supply _Action_Plan.pdf.

Vine, Michael. 2018. "Learning to Feel at Home in the Anthropocene: From State of Emergency to Everyday Experiments in California's Historic Drought." *American Ethnologist* 45 (3): 405–16.

Von Schnitzler, Antina. 2008. "Citizenship Prepaid: Water, Calculability, and Techno-Politics in South Africa." *Journal of Southern African Studies* 34 (4): 899–917.

———. 2013. "Traveling Technologies: Infrastructure, Ethical Regimes, and the Materiality of Politics in South Africa." *Cultural Anthropology* 28 (4): 670–93.

———. 2016. *Democracy's Infrastructure: Techno-Politics and Protest after Apartheid*. Princeton, NJ: Princeton University Press.

Voyles, Traci Brynne. 2015. *Wastelanding : Legacies of Uranium Mining in Navajo Country*. Minneapolis: University of Minnesota Press.

Wachsmuth, David, and Hilary Angelo. 2018. "Green and Gray: New Ideologies of Nature in Urban Sustainability Policy." *Annals of the American Association of Geographers* 108 (4): 1038–56.

Wakefield, Stephanie. 2020. "Making Nature into Infrastructure: The Construction of Oysters as a Risk Management Solution in New York City." *Environment and Planning E: Nature and Space* 3 (3): 761–85.

Wakefield, Stephanie, and Bruce Braun. 2019 "Oystertecture: Infrastructure, Profanation and the Sacred Figure of the Human." In *Infrastructure, Environment, and Life in the Anthropocene*, edited by Kregg Hetherington, 193–215. Durham, NC: Duke University Press.

Walker, Jeremy, and Melinda Cooper. 2011. "Genealogies of Resilience From Systems Ecology to the Political Economy of Crisis Adaptation." *Security Dialogue* 42 (2): 143–60.

Walsh, Casey. 2022. "Beyond Rules and Norms: Heterogeneity, Ubiquity, and Visibility of Groundwaters." *Wiley Interdisciplinary Reviews: Water* 9 (4): e1597.

Watkins, Claire Vaye. 2015. *Gold Fame Citrus*. New York: Riverhead Books.

Watts, Michael. 2015. "Now and Then: The Origins of Political Ecology and the Rebirth of Adaptation as a Form of Thought." In *The Routledge Handbook of Political Ecology*, edited by Tom Perreault, Gavin Bridge, and James McCarthy, 41–72. London: Routledge.

Weheliye, Alexander. 2014. *Habeas Viscus: Racializing Assemblages, Biopolitics, and Black Feminist Theories of the Human*. Durham, NC: Duke University Press.

Weizman, Eyal. 2002. "The Politics of Verticality." Open Democracy. http://www.opendemocracy.net/ecology-politicsverticality/article_801.jsp.

Weszkalnys, Gisa. 2015. "Geology, Potentiality, Speculation: On the Indeterminacy of First Oil." *Cultural Anthropology* 30 (4): 611–39.

White, Richard. 1996a. "'Are You an Environmentalist or Do You Work for a Living?': Work and Nature." In *Uncommon Ground: Rethinking the Human Place in Nature*, edited by William Cronon, 171–85. New York: W. W. Norton.

———. 1996b. *The Organic Machine: The Remaking of the Columbia River*. New York: Macmillan.

Willems, Jannes J., Astrid Molenveld, William Voorberg, and Geert Brinkman. 2020. "Diverging Ambitions and Instruments for Citizen Participation across Different Stages in Green Infrastructure Projects." *Urban Planning* 5 (1): 22–32.

Williams, Joanna. 2019. "Circular Cities." *Urban Studies* 56 (13): 2746–62.

Williams, Joe. 2018a. "Assembling the Water Factory: Seawater Desalination and the Techno-Politics of Water Privatisation in the San Diego-Tijuana Metropolitan Region." *Geoforum* 93: 32–39.

———. 2018b. "Diversification or Loading Order? Divergent Water-Energy Politics and the Contradictions of Desalination in Southern California." *Water Alternatives* 11 (3): 847–65.

Williams, Raymond. 1975. *The Country and the City*. New York: Oxford University Press.

Woelfle-Erskine, Cleo. 2015. "Emerging Cultural Waterscapes in California Cities Connect Rain to Taps and Drains to Gardens." In *Sustainable Water: Challenges and Solutions from California*, edited by Allison Lassiter, 317–41. Oakland: University of California Press.

Woelfle-Erskine, Cleo, July Oskar Cole, Laura Allen, and Annie Danger, eds. 2006. *Dam Nation: Dispatches from the Water Underground*. Portland, OR Soft Skull Press.

Wolch, Jennifer R., Jason Byrne, and Joshua P. Newell. 2014. "Urban Green Space, Public Health, and Environmental Justice: The Challenge of Making Cities 'Just Green Enough.'" *Landscape and Urban Planning* 125:234–44.

Wynter, Sylvia. 1992. "Beyond the Categories of the Master Conception: The Counter-Doctrine of Jamesian Poesis." In *C. L. R. James' Caribbean*, edited by Henry Paget and Paul Buhle, 63–91. Durham, NC: Duke University Press.

———. 2003. "Unsettling the Coloniality of Being/Power/Truth/Freedom: Towards the Human, After Man, its Overrepresentation—An Argument." *New Centennial Review* 3(3): 257–337.

Zee, Jerry C. 2021. *Continent in Dust: Experiments in a Chinese Weather System*. Oakland, CA: University of California Press.

Zeiderman, Austin. 2016. *Endangered City: The Politics of Security and Risk in Bogota*. Durham, NC: Duke University Press.

Zetland, David. 2009. "The End of Abundance: How Water Bureaucrats Created and Destroyed the Southern California Oasis." *Water Alternatives* 2 (3): 350–69.

Zhang, Amy. 2020. "Circularity and Enclosures: Metabolizing Waste with the Black Soldier Fly." *Cultural Anthropology* 35 (1): 74–103.

———. 2024. *Circular Ecologies: Environmentalism and Waste Politics in Urban China*. Stanford, CA: Stanford University Press.

Zuniga-Teran, Adriana A., Andrea K. Gerlak, Alison D. Elder, and Alexander Tam. 2021. "The Unjust Distribution of Urban Green Infrastructure Is Just the Tip of the Iceberg: A Systematic Review of Place-Based Studies." *Environmental Science & Policy* 126 (December): 234–45.

Index

adaptation work, 7–8, 9, 10, 11, 61, 67, 87, 141, 164, 173. *See also* ecosystem duties
affordable housing, access to, 45–46, 182n19
Allen, Jouett, 106
Allen, Laura, 73–74
Amy (former environmental NGO worker), 158–59
Ana (greywater reuse advocate), 72–73, 76, 86
Anand, Nikhil, 121–22
Andrea (environmental NGO worker), 157
Anne (LADWP staffer), 78
Annual Convention of the American Forestry Association, 98–99
aqueducts: history of LA's water system, 13–16, 15*fig*, 19, 188n22; LA water sources, 5, 6*map*, 189nn29–30. *See also* LA Aqueduct system; LA Department of Water and Power (LADWP)
archival research, 30–31
Ariana (residential green infrastructure adoptee), 24–25, 159
Arleta, 105*map*
Army Corps of Engineers, 102, 107, 176n11

Bakker, Karen, 130–31
Battistoni, Alyssa, 9
Besky, Sarah, 9

Bigger, Patrick, 7
Björkman, Lisa, 43
"blackwater," 73
Blanchette, Alex, 9
Bode, Sam, 55–57
Broadous, Hillary, 107
"bucketless water buckets," 117
building codes, 73
Bureau of Reclamation, 130–31, 134–35
Bureau of Sanitation, City of LA, 37, 51–54, 67, 104, 117, 120, 139–40, 150–54, 192n24

California: climate of, 12–14; history of LA's water system, 13–16, 15*fig*, 19
California Aqueduct, 5, 6*map*
California Department of Health Services, 108–9
California Department of Water Resources (DWR): *Residential Greywater Management* report (1980), 74
California Plumbing Code, 73–76
California State Assembly: AB 3518, residential greywater regulations, 75
California Supreme Court, 14, 104, 189n28
capitalism: ecological work outside of value, 147–49; racial capitalism, 95, 104–11, 105*map*, 109*fig*, 110*map*, 148–49, 182n19

Founded in 1893,
UNIVERSITY OF CALIFORNIA PRESS
publishes bold, progressive books and journals
on topics in the arts, humanities, social sciences,
and natural sciences—with a focus on social
justice issues—that inspire thought and action
among readers worldwide.

The UC PRESS FOUNDATION
raises funds to uphold the press's vital role
as an independent, nonprofit publisher, and
receives philanthropic support from a wide
range of individuals and institutions—and from
committed readers like you. To learn more, visit
ucpress.edu/supportus.

www.ingramcontent.com/pod-product-compliance
Lightning Source LLC
Chambersburg PA
CBHW020856270326
41928CB00006B/728